Minds Before Machines

Minds Before Machines

How Ancient Thought Solved Problems that Modern Tools Can't

Rowan Mercer

Mindful Pages

Published in 2025

ISBN: 9789371774093 (PB)
ISBN: 9789371770415 (eBook)

Published by

Mindful Pages
Imprint of Alpha Editions LLC
312 W. 2nd St #1834
Casper, WY 82601, USA

This book was developed using advanced writing tools to support the drafting and ideation process. All content has been thoughtfully curated, refined, and reviewed to ensure quality and relevance.

Content

Introduction
The Mind as the First Machine

Long before bronze tools gleamed in the sun or the first clay tablets bore lines of script, humans had already harnessed their most potent instrument: the mind. Bereft of wheels, compasses, ploughs, or engines, our ancestors faced the same elemental challenges that continue to confound us today. How does one cross an open sea without sight of land? How can massive stones be lifted without steel or pulleys? How can communities be fed without the benefit of modern chemistry? The solutions they devised did not emerge from external implements but from cultivated habits of thought—rigorous observation, careful reasoning, and the weaving together of collective memory.

This book rests on a deceptively simple premise: the true history of human ingenuity is not found in artefacts but in ways of thinking. Objects reveal what was built; they rarely reveal how minds were prepared to build it. By shifting our gaze from tools to the intellectual frameworks that guided them, we uncover the subtle, often invisible strategies that allowed ancient societies to resolve problems of extraordinary complexity with remarkable clarity.

Surrounded today by a profusion of tools—computers that simulate entire worlds, algorithms sifting oceans of data, devices that collapse distances in seconds—we are easily seduced into believing that modern dilemmas demand ever more sophisticated machines. Yet many of our gravest challenges remain cognitive rather than material. We are swamped by complexity, paralysed by abundance, distracted by noise. The scarcity that once honed ancient minds has given way to a glut that dulls our own. To recover their strategies is not mere nostalgia; it is a return to first principles of human thought under pressure.

The Unseen Machinery of Thought

The metaphor of the machine is apt. Machines transform inputs into outputs through organised processes. The human mind was the first machine of consequence, long before cogs, levers, or engines. It could convert perception into knowledge, knowledge into strategy, strategy into survival. Navigating a featureless ocean demanded a mind capable of discerning patterns in stars, swells, and winds with painstaking attention. Building a temple aligned to the heavens required an imagination that grasped geometry not merely as abstract numbers, but as an echo of cosmic order.

These feats remind us that intelligence is not simply possession of information but mastery in structuring it. Ancient cultures refined systems for compressing and transmitting knowledge—through oral traditions, ritual, and mnemonic techniques—creating intellectual infrastructures every bit as essential as ploughs or spears.

A Book About Thinking, Not Tools

This book is not a catalogue of ancient artefacts. It does not linger on the spectacle of pyramids or marvel at aqueducts for their own sake. Rather, it traces the invisible lines of reasoning beneath them. How did scarcity become a spur to invention rather than an obstacle? How did societies without paper, ink, or blueprints sustain technical knowledge across centuries? Why did philosophers in Greece, India, and China conceive abstract principles that would later underpin mechanics, ethics, and mathematics? These questions concern not what was built, but how thought was disciplined, cultivated, and transformed into action.

History shows that problem-solving is no modern invention. Our assumption that more advanced tools imply superior thinking is misleading. Those who mapped the stars without telescopes, irrigated deserts without pumps, or raised monuments without steel were not less intelligent than we are—they were differently intelligent, shaped by scarcity, constrained by necessity, and compelled to innovate in ways that speak powerfully to our own age of overwhelming abundance.

Why the Past Matters for the Present

Why revisit these ancient modes of thought now? Because our technological power has multiplied, not simplified, the complexity we face. Machines can calculate instantly, yet decisions elude us. We can move mountains, yet struggle to sustain ecosystems. We can transmit data across the globe in milliseconds, yet fail to preserve knowledge across a single generation.

To rediscover how earlier minds approached such challenges is to recover strategies that remain profoundly relevant. The Greeks crossed seas with no maps. The Incas maintained vast agricultural systems without the wheel or written numerals, sustaining populations in landscapes that would challenge modern engineers. Roman builders created concrete structures still standing today. These examples are not quaint curiosities; they are evidence that a disciplined, imaginative mind—sharpened by necessity—can devise solutions of lasting significance.

Beyond Romanticism

This is not a sentimental tale of golden ages. Ancient solutions were imperfect and often fragile. Civilisations collapsed through ecological mismanagement, monuments crumbled from error, wars destroyed societies themselves. Failures are as instructive as triumphs, revealing the limits of thought and the dangers of hubris. Understanding ancient ingenuity requires confronting its fragility as well as its power.

The Architecture of the Book

Each chapter examines a domain of human challenge and the mental strategies developed to meet it. We begin with constraint—how scarcity compelled refinement of ideas. We explore wayfinding, revealing how vast journeys were navigated through stars, waves, and oral cartography.

Architecture, agriculture, warfare, mathematics, medicine, energy, and urban design follow, considered not for their artefacts but for the intellectual scaffolding that made them possible. Later chapters examine knowledge transmission, lessons from failure, and the frameworks of imagination, discipline, and constraint that remain instructive today. The structure is thematic, emphasising patterns of thought that transcend civilisations and centuries.

An Invitation to the Reader

This book is, above all, an invitation. It invites you to consider what the mind, cultivated as the first and most enduring machine, can achieve. It urges you to regard constraint not as hindrance, but as a catalyst for clarity. It asks you to see memory, observation, and philosophy as living technologies of thought, not relics.

You are not here to admire pyramids or aqueducts as static marvels. You are invited into the workshops of ancient thought, to witness how minds shaped the world in ways that continue to resonate. Attend closely, and you will not merely learn history—you will sharpen your own capacity to navigate complexity with insight.

In the end, the promise of this book is clarity. Clarity born from rediscovering how human beings, under scarcity and uncertainty, harnessed imagination, discipline, and collective memory to solve enduring problems. Machines will continue to evolve, technologies will astonish. Yet the strategies of thought that first enabled them are timeless, and remain the truest measure of human ingenuity.

Here, we begin: with the mind as the first machine, and with the conviction that by studying its earliest workings, we may yet learn to think more clearly, act more wisely, and meet the challenges of our own age of complexity.

Chapter 1
Thinking in Constraint

Constraint has long been cast as the enemy of creativity, the blockade that stifles imagination and suppresses invention. Yet history repeatedly tells a different story: limitation can sharpen thought, compel precision, and yield solutions of striking ingenuity. When faced with scarcity—of materials, labour, or tools—ancient innovators did not abandon ambition; they refined it. They learned to think not with abundance, but against its absence. This chapter begins with that paradox: necessity, far from hindering invention, became one of its most potent design principles.

Consider the Egyptian builders who raised colossal monuments without steel scaffolding or powered cranes. Every stone moved, every ramp constructed, demanded calculations performed under austere conditions. The absence of machinery was not emptiness but a problem to be solved, compelling engineers to exploit leverage, geometry, and human muscle in ways both minimalist and monumental. Or consider the Polynesian navigators crossing the Pacific without charts or compasses. Lacking physical instruments, they cultivated acute sensitivity to stars, swells, and winds, transforming their bodies and memories into living tools. These are not tales of deprivation but of adaptation, where constraint became an advantage precisely because it demanded intellectual discipline.

Constraint functions not as a barrier but as a catalyst. Abundance tempts excess, layering solution upon solution.

Scarcity forces harder questions: what is essential? How can less achieve more? How can knowledge substitute for materials? The relentless paring down produced solutions elegant, efficient, and enduring. The Roman aqueduct, carrying water across miles of uneven terrain, exemplifies restraint: gravity, not pumps or endless labour, served as the primary instrument. The genius lay in knowing how to harness what existed.

But constraint alone was insufficient. Observation—disciplined, systematic, and rigorous—was the second pillar of ancient problem-solving. In a modern world rich with instruments extending human perception, it is easy to underestimate the intellectual power of unaided senses. Yet for ancient societies, observation functioned as a technology in its own right. The Babylonians tracked planetary motions with the naked eye, recording patterns over centuries to forecast celestial events. Nile farmers read river swells, paired with the rising of Sirius, to determine planting cycles. Observation was never passive; it demanded attention, repetition, and disciplined interpretation.

The Greeks formalised this into proto-scientific practice. Aristotle, flawed by modern standards yet extraordinary in method, exemplified disciplined observation—cataloguing, reasoning, seeking patterns in the visible world. In China, scholars observed natural cycles, distilling them into principles guiding agriculture, medicine, and governance. Without modern instruments, humans cultivated their senses into finely tuned tools for reading the world.

Observation, however, needed preservation across time. Collective memory emerged as a crucial technology. Before blueprints or codices, societies embedded technical knowledge into oral traditions. Polynesian chants encoded star maps. West African griots maintained genealogies with remarkable precision. Inca quipus—knotted strings—served as tactile numerical records. These were deliberate, engineered systems, ensuring knowledge survived beyond individual lifetimes. Memory, in this sense, was infrastructure: apprenticeships, rituals, and communal recitations acted as living archives, adaptable and resilient.

Even observation and memory required frameworks of reasoning to be fully effective. Philosophy provided this scaffolding. Greek abstraction of mechanics—levers, pulleys, balances—emerged from debates about motion, causality, and purpose. Aristotle's notions of potential and actuality, though imperfect physically, laid foundations for later mechanical thought. Confucian principles in China shaped governance and infrastructure, embedding harmony and hierarchy into both social and physical designs. Roads, canals, and bureaucracies were constructed not merely for utility, but to express rational order.

This dialogue between philosophy and practice was not limited to intellectual elites. Roman engineers, in building aqueducts, relied on philosophical ideas of proportion and order. Indian mathematicians, confronting zero, were reshaping logic itself. Philosophy became a practical tool, a conceptual lever as crucial as any fulcrum in engineering.

This chapter, therefore, is not a catalogue of inventions but an exploration of four intertwined strategies defining ancient problem-solving: necessity, observation, memory, and philosophy. Scarcity fostered precision. Disciplined senses became instruments. Knowledge was a communal infrastructure. Ideas themselves served as tools to shape the world.

Modern readers have much to gain from revisiting these strategies. In an age where abundance often breeds inefficiency, where digital tools overwhelm and data obscures, the lessons of constraint are urgent. How might imposing limits yield more elegant solutions? How might careful observation uncover patterns lost in noise? How might collective memory preserve knowledge beyond the fragility of software or institutions? How might philosophical reflection guide practical choices in ways both effective and humane?

The ancients demonstrate that problem-solving is less a matter of material resources than of mental orientation. Constraint can paralyse or catalyse; it can induce resignation or refinement. For those who laid the foundations of civilisation, constraint most often produced invention. In that alchemy—turning scarcity into a wealth of thought—lies the enduring power of ancient ingenuity.

Necessity as Design Principle

Scarcity is often imagined as a condition to be overcome, a weakness to be corrected, or a hindrance to progress. Yet some of the most enduring achievements of human civilisation were born precisely out of constraint. When resources were scarce, when labour was limited, when tools were crude or entirely absent, ancient innovators responded not by retreating from ambition but by refining it. Their inventions were not the product of excess but of discipline, and it is in this paradox that one finds the deeper logic of human ingenuity: *constraint is not an obstacle to innovation but one of its most powerful design principles.*

The Great Pyramids of Giza, monumental in scale and timeless in endurance, were raised without cranes, iron machinery, or the logistical capacities of modern industry. The builders were compelled to think in terms of leverage, balance, and coordination rather than brute force. Wooden sledges dragged blocks across sand lubricated with water, ramps inclined at precise angles, and counterweights distributed human labour with efficiency. Each of these techniques was not merely a workaround for missing technologies but a system deliberately honed by scarcity. The absence of abundance forced design choices that prioritised elegance over extravagance. Similarly, the aqueducts of Rome, delivering fresh water over vast distances, relied upon gravity as their engine. Pumps existed in crude forms, yet the decision to use gradients and arches was not simply practical but profoundly economical. It required the Roman engineer to imagine constraint not as a limit but as a structural principle around which the entire design revolved.

In many cases, the very scarcity of resources spurred cultures to transform intellectual effort into a resource in itself. Ancient Greek mathematicians, confronted with limited instruments, refined deductive geometry into a tool capable of solving practical engineering problems without a single physical prototype. Euclidean logic became a way of modelling and testing structures before a stone was ever laid. By contrast, cultures with abundant natural resources often indulged in overbuilt, inefficient designs that collapsed under their own weight. The delicate terracing systems of the Andes, crafted by the Inca, demonstrate this vividly. Farming on steep mountain slopes without abundant arable land or advanced machinery demanded a solution that could never rely on scale alone. The terraces not only prevented erosion but created microclimates for crops. Each stone placed was an act of economy, where knowledge substituted for excess. It is striking that such systems remain viable and sustainable even today, long after modern mechanised agriculture has eroded soil at alarming rates.

The principle of necessity shaping design extended beyond physical structures into the organisation of labour and the management of societies. In Mesopotamia, where fertile land existed but water was unpredictable, scarcity of reliable harvests produced irrigation systems of remarkable sophistication. Canals, gates, and dykes were engineered not with surplus steel but with mud brick and human calculation. The limitation of materials forced a deeper engagement with hydrology, leading to a system that was as much intellectual as physical. It was not wealth but need that taught Mesopotamian farmers to read the language of rivers.

Constraint also forged cultural values around frugality and efficiency. In Japan, the principle later known as *ma*—the idea of purposeful emptiness—can be traced back to resource-limited environments where design was compelled to use space, timber, and stone sparingly. Though emerging much later in history, this philosophy reveals a pattern that repeats across civilisations: when abundance is absent, design becomes an act of choosing what not to use. *Elegance is the child of scarcity.*

One of the most powerful aspects of necessity as a design principle is its role in preventing excess complexity. Modern projects often collapse under the weight of their own abundance: too many moving parts, too many redundancies, too much dependence on fragile supply chains. By contrast, ancient innovation was necessarily pared down. The Polynesian canoe was not an ornate construction but a model of distilled function. Every fibre of wood, lash of rope, and strip of sail had to justify its place because resources were limited and the stakes of failure—on open ocean—were absolute. Constraint demanded that every design element was essential, and in that demand lay both safety and durability.

This does not mean scarcity was universally beneficial. There were times when constraints led to stagnation, when the absence of resources paralysed rather than sharpened. But the critical difference lies in whether societies treated scarcity as a barrier or as a framework. Where imagination was channelled through necessity, innovation flourished. Where imagination was stifled by the weight of limitation, collapse followed. It is here that we see the true intellectual dimension of constraint: it is not simply about having less, but about rethinking what *less* makes possible.

Modern design philosophies, particularly those influenced by systems thinking and lean manufacturing, have rediscovered this ancient insight. Engineers speak of "designing for constraint" as a method to strip away excess, focusing on what is essential and adaptable. Start-ups in technology sectors deliberately impose resource limits to force creativity, mirroring the very conditions under which Polynesian navigators, Mesopotamian farmers, and Incan terrace builders once operated. The echo across millennia is clear: when excess is removed, clarity emerges.

Yet there is a philosophical tension worth recognising. Abundance promises freedom, yet it often produces dependence—on resources, on tools, on complex systems whose fragility is hidden until crisis strikes. Scarcity restricts, yet it often produces resilience—systems designed under constraint are robust because they have been tested against the margins of failure from the start. The Roman aqueduct that still stands, the Incan terrace that still holds soil, the Polynesian canoe remembered in oral tradition—all testify to this paradox. *Constraint is not only a teacher of ingenuity; it is a guarantor of durability.*

To read history with this lens is to shift admiration from the finished object to the intellectual process behind it. A pyramid is not merely a pile of stone; it is a crystallised lesson in how to think under limitation. An aqueduct is not just arches across a valley; it is a reminder that when pumps were absent, minds reimagined water as an obedient force of gravity. The terraces of the Andes are not simply agricultural curiosities; they are living proof that design refined by necessity can outlast empires.

In our own age, surrounded by unprecedented abundance, these lessons have particular urgency. We live in systems that often assume unlimited resources—cheap energy, endless raw materials, infinite computational capacity. But the fragility of these assumptions is revealed in every supply chain disruption, every energy crisis, every ecological collapse. What ancient innovators teach us is not to romanticise scarcity but to recognise its hidden gift: clarity of purpose. When every resource must be accounted for, when every design choice carries weight, when every mistake risks collapse, ingenuity ceases to be optional. It becomes the condition of survival.

The idea that necessity functions as a design principle is therefore not a historical curiosity but a living insight. It explains why some of the greatest achievements of the past remain benchmarks of durability, efficiency, and elegance. More importantly, it challenges us to reconsider abundance not as an unqualified advantage but as a condition that often obscures clarity. Constraint compels precision. Scarcity forces imagination. And in the discipline of thinking with less, human beings have often discovered more enduring solutions than in any age of excess.

Observation as Technology

The most powerful instrument available to the ancient world was not fashioned from metal, wood, or stone; it was the human capacity to observe. Long before glass lenses magnified the stars or microscopes revealed the invisible, entire civilisations relied on the disciplined use of the senses to identify, refine, and transmit patterns in nature. Observation was not casual noticing but cultivated attention, trained through repetition and collective practice until it functioned as a technology in its own right. It was through such disciplined seeing and listening that ancient peoples learned to cross oceans without maps, harvest crops in rhythm with the cosmos, and forecast weather long before the first barometer was built. *Observation, at its most disciplined, became the first universal instrument of human civilisation.*

Few examples illustrate this more vividly than the navigators of Polynesia. Spread across thousands of islands and separated by expanses of ocean so vast that even European sailors centuries later considered them untraversable, the Polynesians developed a system of navigation that relied entirely on observation. By watching the rising and setting of stars, noting subtle shifts in the colour and texture of ocean swells, and recognising the flight paths of seabirds, they could locate land hundreds of miles beyond the horizon. This was not instinct but trained perception: years of apprenticeship under master navigators, who taught through chants, stories, and demonstrations. To speak of their voyages as accidental "drift" is to ignore the intellectual infrastructure behind them. They had transformed their senses into calibrated instruments, capable of reading information from environments others regarded as featureless void. The discipline of observation became, quite literally, a tool for survival.

Agricultural societies, no less than seafaring ones, depended upon the ability to read patterns invisible to the untrained eye. The Nile flood cycle was synchronised with the heliacal rising of Sirius, a star whose reappearance at dawn became the earliest marker of the Egyptian calendar. Farmers in Mesopotamia watched for subtle changes in the Tigris and Euphrates to anticipate irrigation needs. The Mayan civilisation tracked celestial bodies with such care that their calendars rivalled modern astronomical calculations in accuracy. These were not the result of isolated flashes of insight but of cumulative observation across generations, compiled, refined, and trusted. *Where writing was absent, memory held; where instruments were absent, the body became the lens.*

Observation also functioned as a way of compressing complexity into usable knowledge. Consider the Andean farmers who worked in dramatically varied microclimates, where altitude could change temperature and moisture levels within a single valley. By monitoring which plants thrived in different niches, they developed systems of vertical agriculture that maximised biodiversity and resilience. The observation was not only of plants but of animal behaviour, cloud patterns, and soil colour, each read as a sign. In such systems, knowledge was not theoretical but practical, rooted in long attention to the subtleties of an environment. The limitation of tools forced deeper reliance on what the senses could register, and that reliance created knowledge systems of striking sophistication.

Yet to describe observation as merely practical understates its philosophical weight. The act of disciplined seeing has always been bound to the search for order. The Greeks codified this impulse, turning empirical noticing into a proto-scientific methodology. Aristotle's catalogues of animals, Theophrastus' studies of plants, and later Ptolemy's astronomical records all illustrate how observation could serve as a foundation for broader theories about causality, order, and the structure of the cosmos. In China, Confucian and Daoist traditions incorporated observation into moral and political philosophy, interpreting the natural world as a guide to human order. Cycles of seasons were not only agricultural cues but metaphors for governance, health, and balance. Here, the line between observation and philosophy blurred; to observe was not merely to notice but to interpret, to make sense of the world by discerning regularities that could be trusted as guides for human action.

What distinguished observation as a technology from casual awareness was its collective discipline. A single farmer may notice the behaviour of birds before rain, but it is the repeated confirmation of many observers across time that transforms such noticing into reliable knowledge. Oral traditions became vehicles for transmitting these observations, embedding them in ritual, story, and instruction. In Aboriginal Australian songlines, patterns of the land are encoded in stories that function simultaneously as cultural heritage and navigational guide. In medieval Europe, weather proverbs ("Red sky at night, shepherd's delight") condensed centuries of observation into phrases memorable enough to be passed down. Such expressions are often dismissed as folklore, yet they represent a distillation of empirical noticing into a form robust enough to survive across generations. *The transmission of observation was itself a kind of technology, ensuring that no single lifetime's attentiveness was wasted.*

There is also a critical contrast worth acknowledging: while observation enabled predictive capacity, it was never infallible. The reliance on celestial events for agriculture worked in regions where climate patterns were stable, but in times of disruption—volcanic eruptions, prolonged droughts, or shifting river courses—observation could mislead. The Mayans, for all their astronomical precision, could not prevent ecological collapse exacerbated by overexploitation of land. Observation sharpened perception but could not eliminate risk. Its power lay not in guaranteeing certainty but in enhancing the odds of survival. This recognition tempers any romanticism: the disciplined senses were extraordinary instruments, but they operated within the fragile limits of the natural world.

In many ways, modern science is a direct heir to this tradition of disciplined observation, though mediated now by instruments that extend our senses. Telescopes, spectrometers, satellites, and sensors all automate and refine what once required direct human attentiveness. Yet the principle remains unchanged: knowledge begins with noticing. A climate scientist reading long-term satellite data is engaged in a form of pattern recognition that echoes the farmer watching flood cycles. The difference is scale and precision, not kind. The continuity reminds us that observation is not a primitive precursor to "real" technology but a technology in itself, one that remains foundational even in the most advanced sciences.

In practical terms, the ancient reliance on observation offers insights still relevant to contemporary challenges. In medicine, for instance, while diagnostic machines dominate, the skill of a physician's eye—spotting the subtle signs of jaundice, the faint tremor of a limb—remains indispensable. In ecological monitoring, satellite data is invaluable, but local communities who observe shifts in animal migration or flowering seasons often provide the earliest warnings of environmental change. Even in fields such as finance or geopolitics, professionals are valued not merely for access to data but for their capacity to discern patterns others overlook. *The discipline of observation, whether of stars, crops, or markets, remains a competitive advantage in a world flooded with information but starved of clarity.*

Perhaps the deepest lesson is that observation demands humility. To attend carefully is to acknowledge that the world discloses its patterns slowly, requiring patience and deference. Ancient societies understood this intuitively: the stars would reveal themselves only through years of watching; the soil would yield its secrets only through repeated planting. In our time, when immediacy and automation tempt us to overlook the subtle, the discipline of observation offers a corrective. It reminds us that knowledge is not only what can be computed but what can be patiently seen, heard, and remembered.

Observation as technology was, and remains, a testament to the ingenuity of human beings when deprived of external instruments. It transformed the senses into tools of extraordinary reliability, capable of guiding ships, sustaining agriculture, and anchoring philosophy. Its lessons are as relevant today as they were thousands of years ago: that attention, when disciplined, is not passive but generative; that patterns, when recognised, turn uncertainty into guidance; and that the simplest instrument of all—the human act of noticing—remains indispensable to understanding and navigating the complexities of the world.

Collective Memory and Oral Engineering

The oldest blueprint is not ink on papyrus or etching on clay but a story, a chant, or a gesture repeated until it became reliable. Before the invention of writing, and even long after script existed but remained restricted to elites, human societies built vast infrastructures of knowledge without recourse to paper or plans. They engineered memory. Oral traditions did more than preserve myths and genealogies; they carried technical instructions, construction methods, navigational routes, agricultural cycles, and entire systems of governance. *Collective memory functioned as an invisible archive, engineered deliberately to be portable, durable, and trustworthy across generations.*

Nowhere is this more striking than in the Polynesian voyaging traditions. Navigators who sailed thousands of kilometres across the Pacific, connecting scattered islands over an ocean that covers a third of the planet, carried no maps, charts, or compasses. Instead, they relied upon complex oral systems: chants encoding the order of star risings, stories embedding the swell patterns of particular seas, and genealogical recitations that doubled as navigational sequences. These were not casual tales but structured technologies of recall, designed with rhythm, redundancy, and symbolism to lock vast quantities of data into memory. Apprenticeship under a master navigator was less about rote memorisation and more about embodying an entire epistemology—how to think with the sky, the sea, and the human voice as instruments. Oral engineering allowed the Pacific to be crossed with confidence centuries before European ships dared venture far beyond sight of land.

Agricultural societies deployed similar systems to encode seasonal knowledge. Among the Dogon of Mali, for instance, oral traditions contained astronomical observations—such as detailed accounts of the cycles of Sirius—that remain debated for their precision. Whether every detail reflects empirical accuracy or symbolic layering, the key point is that oral frameworks allowed communities to store and transmit astronomical knowledge integral to agriculture and ritual. In India, the Vedic corpus was preserved with astonishing fidelity for centuries through oral recitation, using elaborate mnemonic devices such as overlapping sequences, pitch modulation, and redundancy patterns. The Rigveda, consisting of over 1,000 hymns, survived intact because its oral transmission was treated as an engineering problem: how to prevent entropy in the human archive. The solution was to make memory collective, distributing responsibility across groups, embedding redundancy into practice, and tying fidelity to ritual significance. *By transforming memory into communal performance, societies achieved a level of durability comparable to stone inscriptions, without ever lifting a chisel.*

Memory was not simply about storage; it was about precision under constraint. Consider the Inca quipu: bundles of knotted strings that encoded numerical data, from census figures to agricultural tallies. Though not oral in the strictest sense, the quipu represents an extension of the same principle—knowledge made portable, reliable, and interpretable without script. Quipucamayocs, the specialists trained to read and transmit these devices, did not operate in isolation; their skill was validated by communal cross-checking, ensuring accuracy through redundancy. The quipu, combined with oral explanation, became a distributed record system that allowed the Inca to administer an empire across mountainous terrain without written language. To call this primitive is to misunderstand its sophistication. It was not a failure to invent writing but a choice to optimise memory into a system fit for its context.

What made these systems effective was their engineering character. Oral traditions were not left to chance but designed with mechanisms that mirrored the redundancies of modern data storage. Repetition created resilience. Song embedded technical detail in rhythm, making it easier to recall and harder to corrupt. Story encoded information in narrative form, providing hooks for memory. Ritual performance acted as periodic checksums, ensuring accuracy through repeated, communal verification. These strategies reveal a profound insight: when the medium is fragile, the system must be robust. Human memory is fallible, but *collective memory, engineered through form and practice, can approach the reliability of written archives.*

The philosophical significance of oral engineering is equally important. Knowledge in oral cultures was not divorced from its social context; it was always relational. A builder did not consult an impersonal manual but learned through apprenticeship, embodying knowledge in practice and memory. This created a deep interdependence between individual skill and communal responsibility. Forgetting was not merely a personal lapse; it was a social failure. By embedding technical knowledge in collective performance, societies ensured that the burden of preservation was shared. In this sense, oral engineering was not just a substitute for writing—it was a fundamentally different epistemology, one in which memory was not externalised into objects but internalised into community.

This reliance on collective memory shaped the very nature of innovation. In literate societies, invention often appears as rupture: new blueprints replace old, written treatises displace earlier models. In oral societies, change was more incremental, woven into existing forms without discarding them. A new navigational insight might be folded into a chant, a new farming technique embedded into a proverb. Innovation was absorbed rather than archived. The system resisted obsolescence because it could not afford obsolescence. Its strength lay in adaptability: as long as the chant was sung, the knowledge remained alive, responsive to change, and bound to practice.

Critically, oral engineering was not infallible. Knowledge could be lost through disruption of transmission—if elders died without passing on chants, if communities were scattered, if ritual structures collapsed. Colonisation and forced assimilation often targeted these systems, recognising that breaking oral continuity dismantled the archive itself. Yet the resilience of many oral traditions even under such pressure testifies to their durability. Aboriginal Australian songlines, for instance, continue to preserve geographical and ecological knowledge across tens of thousands of years, guiding movement, seasonal hunting, and social organisation. Such persistence demonstrates that oral engineering, far from fragile, can exceed the longevity of many written records. Stone tablets crack; archives burn; but a song carried by living voices can outlast empires.

In modern contexts, the lessons of oral engineering remain pertinent. Organisations today face the challenge of knowledge management in environments saturated with information yet vulnerable to loss. Databases become obsolete, digital formats expire, and institutional memory is fragmented by turnover. The ancient insight that memory must be engineered collectively—through redundancy, performance, and embedding knowledge in practice—offers practical guidance. Corporations that cultivate mentorship, apprenticeships, and narrative-based knowledge sharing are, in effect, replicating strategies once used by navigators, farmers, and priests. *When knowledge matters most, it must be made social, not left solely to documents or devices.*

What stands out most is that oral engineering required not less discipline than writing but more. The absence of externalised records demanded vigilance, repetition, and precision at levels modern societies often assume unnecessary. Yet that very vigilance honed capacities of memory and attention that literate and digital cultures risk losing. The Greek rhapsode capable of reciting Homer's epics, the Vedic priest preserving thousands of verses through voice alone, the Polynesian navigator carrying the map of the stars in his body—these figures represent not curiosities but reminders of what the human mind, collectively disciplined, is capable of sustaining.

Collective memory and oral engineering demonstrate that the human archive need not be written to be powerful. When cultivated through design principles of rhythm, redundancy, and ritual, memory can carry the weight of civilisations. These systems remind us that knowledge is not only what is stored in texts or devices but what is embedded in communities, rehearsed in practice, and safeguarded through collective discipline. Far from being fragile placeholders awaiting the arrival of writing, oral traditions were robust, adaptive, and astonishingly effective. They reveal a truth easily forgotten in a world enamoured with external tools: that the most enduring repositories of knowledge may not be housed in libraries or servers but in the trained, remembering mind of a community determined never to let its knowledge vanish.

Philosophy as Engineering Logic

The bridges, aqueducts, and temples of antiquity did not rise from raw trial and error alone. They were shaped by frameworks of thought—abstract ideas that provided structure long before blueprints or schematics existed. Philosophy, in its earliest and most practical sense, was not an indulgence of speculation but a form of engineering logic. It was a way of organising the mind so that the chaos of materials, labour, and ambition could be translated into order. *Behind every enduring structure lay not only stone and timber, but a set of principles about balance, proportion, and purpose derived from philosophical reasoning.*

Aristotle is often remembered for metaphysics and ethics, but his contributions to mechanics illustrate how philosophy translated directly into practical design. His distinction between potential and actuality, between forces latent and forces expressed, became the conceptual soil in which later mechanics grew. Though his theories were imperfect—he misunderstood inertia, for instance—they provided a framework for thinking systematically about cause and effect. The notion that every movement requires explanation, that forces and counterforces operate in patterned ways, encouraged engineers to look for underlying principles rather than accept outcomes as arbitrary. In this sense, Aristotle's philosophy of motion was less important for its accuracy than for its insistence that practical problems could be reduced to logical inquiry. When a builder raised a lever or balanced a beam, they were applying principles that philosophy had helped to render thinkable.

In China, philosophy shaped infrastructure not through mechanics but through order. Confucianism's emphasis on harmony, hierarchy, and moral rectitude extended into governance and civic design. The grid layouts of Chinese cities, with their central axes and hierarchically arranged spaces, were not merely efficient but expressions of a cosmological and ethical order. Roads, canals, and administrative systems were aligned with ideas about balance between Heaven, Earth, and humanity. The same logic that dictated familial respect and social responsibility also dictated that water must flow in regulated channels, that streets must converge on central authority, and that ritual spaces must be distinct from markets. *Philosophy in this context was not decoration to engineering; it was the logic that determined engineering choices themselves.*

The Greeks and Chinese were not alone in embedding philosophy into practice. In India, the concept of *dharma*—order, duty, and right action—infused both legal systems and technical thought. The Sulbasutras, ancient Indian texts on altar construction, combined geometric insight with ritual precision. They reveal an intellectual environment in which abstract reasoning about proportion and harmony was inseparable from religious and social obligation. Constructing a fire altar was not just a technical feat; it was a manifestation of cosmic principles. The geometry was practical, but it was also philosophical—a demonstration of how human action aligned with the deeper order of the universe.

What unites these traditions is the recognition that philosophy provided scaffolding for technical reasoning. Abstract categories—cause and effect, harmony and disharmony, proportion and imbalance—served as mental templates applied to material problems. This was particularly crucial in the absence of detailed written manuals or precise measurement tools. When resources for technical instruction were limited, philosophy supplied a shared vocabulary that allowed groups to conceptualise, debate, and refine solutions. It gave shape to intuition, directing it towards systematic practice.

This does not mean philosophy was always an asset. At times, abstract systems ossified into dogma that constrained innovation. Aristotle's mistaken physics, treated as authoritative for centuries, slowed progress in mechanics until empiricism corrected it. In China, Confucian orthodoxy sometimes stifled more experimental approaches, discouraging departures from established order. Yet even here, the point is not that philosophy impeded practice, but that its influence was inescapable. The very logic that constrained could also enable, depending on how it was applied. *Philosophy was never neutral; it set the terms within which problems could be conceived and solutions pursued.*

The relationship between philosophy and engineering logic is visible not only in monumental works but in the subtler organisation of labour and knowledge. Roman engineering, for instance, was grounded in a Stoic appreciation for rational order and endurance. The emphasis on universality and lawfulness mirrored the way Roman architects standardised construction techniques, replicating aqueducts and roads across diverse terrains. Their trust in system over improvisation reflected a philosophical worldview in which nature was rational and could be harnessed through disciplined order. The enduring strength of Roman infrastructure speaks not only to technical skill but to the philosophical conviction that systems could be both rational and replicable.

In the Islamic Golden Age, thinkers such as Alhazen (Ibn al-Haytham) demonstrated how philosophical commitment to reason and scepticism could guide practical experimentation. His insistence on testing ideas through observation rather than relying solely on authority was a philosophical stance with direct engineering implications, particularly in optics and architecture. Here philosophy functioned as a critical corrective, pushing practice away from rote repetition and towards methodological innovation. It is telling that his *Book of Optics* influenced both medieval science and practical design, from lens-making to the construction of domes and arches.

Modern readers might assume that engineering now stands apart from philosophy, that the proliferation of technical manuals and computer simulations renders abstract frameworks obsolete. Yet the persistence of design principles such as efficiency, sustainability, and resilience shows otherwise. These are not technical specifications but philosophical commitments, guiding engineers to privilege certain outcomes over others. The language of "human-centred design" or "systems thinking" is, at its core, philosophical, shaping how problems are defined before a single calculation is made. The ancients remind us that philosophy is not an optional luxury but the invisible framework within which all technical practice unfolds.

There is also a cautionary lesson. When philosophy and engineering are divorced, practice risks becoming shortsighted. A society capable of extraordinary technical feats may still fail if its projects lack guiding principles. The collapse of monuments through overreach, the ecological failures of irrigation empires, the moral contradictions of technologies of war—all remind us that *technical capability without philosophical grounding is unstable*. Conversely, when philosophical principles are too rigid, they can freeze practice, preventing adaptation to new challenges. The balance between abstraction and application is delicate but essential.

To study philosophy as engineering logic, then, is to recognise that human progress has always depended on more than clever tools. It has required abstract reasoning to discipline, direct, and sometimes restrain technical ambition. Whether it is Aristotle theorising mechanics, Confucians designing cities as embodiments of order, or Indian mathematicians aligning geometry with ritual, the lesson is consistent: ideas themselves are tools. They frame reality, determine what counts as a problem, and provide the criteria by which solutions are judged. To ignore this is to miss the hidden architecture beneath every aqueduct, temple, or grid of streets.

For the modern reader, the relevance is immediate. We inhabit a world awash with technologies capable of reshaping ecosystems, economies, and even human biology. The question is not whether we can build, but how and why we should. Ancient innovators remind us that philosophy was never separate from engineering; it was its foundation. To recover that insight is to accept that every act of design is also an act of thought, and that the most enduring works are those in which technical ingenuity and philosophical clarity meet.

Chapter 2
Wayfinding Without Maps

Long before ink traced coastlines on paper, humans traversed immense distances guided solely by their minds, their senses, and the rhythms of the natural world. To the modern observer, accustomed to GPS and printed charts, crossing thousands of miles of open sea without instruments seems almost reckless. Yet entire civilisations thrived by cultivating a different art of navigation—one that depended not on tools but on the disciplined organisation of perception and memory. Wayfinding without maps was less instinctual and more a rigorous intellectual practice, transforming stars, waves, and wind into a system of coordinates, embedding routes into memory with a precision that rivalled any chart centuries before cartography existed.

The Pacific Ocean offers perhaps the most extraordinary evidence of this human capacity. Polynesian navigators sailed between scattered islands, distances greater than those separating continents, guided not by compasses but by what might be called mental astronomy. The night sky became a moving map: each star memorised in relation to the horizon, its rising and setting points forming invisible bearings. Navigators carried in their minds structured grids of celestial markers, each linked to direction, distance, and season. The sky itself functioned as a chart, portable, inexhaustible. Apprentices trained for years to internalise these star paths, reciting them in chants until the sequences became instinctive. Patience, discipline, and collective memory were required—but the rewards were voyages across expanses that centuries later often confounded

European explorers, who risked shipwreck or disorientation.

Stars were only part of the system. The ocean provided a second, equally sophisticated compass. Pacific Islanders learned to read swells and currents, discerning subtle cross-waves generated when large swells refracted around islands far beyond sight. A trained navigator, seated in a canoe, could feel the rhythm of water against the hull and know an island lay days away in a particular direction. Winds, shifting predictably with season and latitude, offered further guidance. These methods may sound romantic, yet they were empirical—tested through generations of careful observation. Where modern sailors rely on instruments, ancient voyagers cultivated their bodies as instruments, attuned to information that to the unpractised eye and ear seemed mere noise.

Perception alone was insufficient. These systems depended on memory engineered for durability. Oral cartography transformed ephemeral observation into structured, repeatable knowledge. Routes were encoded in songs and stories, with each landmark, star path, and ocean swell linked to narrative or rhythm. A chant might guide a sailor from one island to another or recall the sequence of landfalls along a trading route. Unlike fragile written maps, these oral systems were communal: rehearsed in ritual, embedded in culture, resilient to loss. They were living maps, where the voice served not merely as expression but as vessel of geography.

The brilliance of these mental maps becomes evident only when contrasted with their limits. Clouds could obscure stars, storms could scramble swells, and misremembered sequences could send ships off course. The history of ancient seafaring includes both triumphs and tragedies—shipwrecks, vanished fleets, migrations ending in loss. These failures underscore the fragility inherent in wayfinding without instruments. Crossing oceans relied not only on courage but on disciplined training; intuition alone was insufficient. The intellectual achievement of ancient navigation is remarkable precisely because it existed in tension with uncertainty.

Wayfinding without maps was not confined to the Pacific. Greek sailors navigated the Mediterranean using constellations, timing journeys by the rising and setting of stars. Arab mariners across the Indian Ocean developed sophisticated systems for reading both stars and seasonal monsoon winds, blending astronomical observation with practical seamanship. What distinguished these cultures was not the invention of instruments—those arrived later—but the cultivation of disciplined perception and abstraction. Observation became systematic, embedded within culture and philosophy, making navigation as much an intellectual tradition as a practical skill.

The philosophical implications are profound. To navigate without maps is to rely on patterns not tangible, to trust abstractions derived from careful perception. Each successful voyage demonstrated that nature could be read and understood without written guides or mechanical aids. Knowledge here blurred the boundary between science and art, practice and philosophy, embedding technical skill within song, story, and ritual.

For modern readers, accustomed to digital navigation, these methods are more than historical curiosities. They invite reflection on thinking without overreliance on tools, on cultivating the mind as an instrument. In an age dominated by GPS and algorithms, the memory of Polynesian navigators or Arab seafarers reminds us of the potential of human perception stretched to its limits. It also serves as a caution: skills unused atrophy; capacities outsourced to machines grow fragile. Ancient wayfinding demonstrates that disciplined attention, embodied knowledge, and communal memory can solve problems even sophisticated technology sometimes cannot.

This chapter explores wayfinding without maps as both technical and intellectual achievement. It examines how cultures used stars, waves, winds, and oral traditions to build navigational systems of remarkable precision and resilience, and it acknowledges the inherent risks and failures. In doing so, it reveals not only the ingenuity of ancient navigators but the cognitive strategies that remain relevant in managing uncertainty today. Crossing an ocean without charts demanded more than bravery—it required a framework of thought: a fusion of perception, memory, and abstraction that continues to offer lessons for navigating a complex, often unmapped world.

Stars as Coordinates

Human beings have always looked up at the night sky, but not all gazes are equal. For some cultures, the stars were not distant ornaments but working instruments—fixed points of reference, carefully catalogued and memorised, that transformed the vault of heaven into a map. To navigate by stars is to treat the cosmos itself as a coordinate system, and what appears effortless in retrospect required disciplined observation, communal memory, and intellectual abstraction. Long before the invention of compasses or sextants, navigators in Polynesia, the Mediterranean, and the Indian Ocean developed forms of *mental astronomy* so refined that they rivalled, and sometimes exceeded, the accuracy of later mechanical devices. *The sky was not symbolic; it was functional. It carried within it a practical geometry that made the impossible voyage thinkable.*

Polynesian voyaging traditions offer one of the most extraordinary cases. Stretching across an ocean that covers a third of the Earth's surface, Polynesian navigators connected islands separated by thousands of kilometres, some no more than specks of land barely visible from sea level. They achieved this not by drifting or by chance, as nineteenth-century Western anthropologists sometimes claimed, but by cultivating star maps memorised and rehearsed as oral knowledge. The principle was elegantly simple but cognitively demanding: each star was associated with a rising and setting point on the horizon, and these points served as bearings. By learning which stars rose directly over specific islands or aligned with particular directions, navigators built mental compasses that functioned with extraordinary accuracy.

Training was long and arduous. A master navigator, or *pwo,* might spend decades teaching apprentices to internalise sequences of stars. Chants, often structured rhythmically to aid recall, served as catalogues of these stellar pathways. A voyage across hundreds of kilometres could be plotted as a sequence: steer towards this star until dawn, then shift to another at its rising, adjust again when a certain constellation tilts. This was not instinct but deliberate cognitive architecture, supported by oral traditions engineered for precision. The feat was intellectual as much as physical. To steer a canoe across the open Pacific was to carry an entire astronomy in one's mind, tested against the horizon each night.

The Greeks, sailing a far smaller but no less challenging sea, developed their own systems of stellar navigation. Homeric epics mention mariners keeping Orion or the Pleiades in sight, while Hesiod advised farmers to plant and harvest according to stellar risings. By the Classical period, Greek sailors had learned to use the North Star as a fixed point. Though Polaris was not perfectly aligned with true north in antiquity, it was close enough to provide orientation. The philosophical curiosity of the Greeks reinforced these practical needs. Astronomers like Hipparchus and Ptolemy catalogued stars systematically, turning observation into ordered knowledge. For the sailor, this meant that navigation could draw upon a growing corpus of astronomical theory. Though their maps were rudimentary compared to modern standards, the intellectual leap was significant: the sky was no longer only a source of myth but a framework for geometry, measurement, and orientation.

Arab navigators in the Indian Ocean carried this fusion of practice and abstraction even further. Building upon both Greek astronomy and indigenous traditions, they created some of the most sophisticated systems of stellar navigation in the premodern world. By the ninth century, Muslim scholars had translated and expanded upon Ptolemaic texts, producing detailed star charts. Mariners applied this knowledge pragmatically, using the rising and setting of certain stars to determine latitude. The *Kamal,* a simple device consisting of a board and knotted string, allowed sailors to fix a star's altitude above the horizon, translating observation into measurable coordinates. Yet behind the device lay a deeper cognitive system: the understanding that consistent angular distances could serve as geographical markers. This was not mere memorisation but the application of an abstract principle—geometry—to practical navigation. The star, in this sense, became both symbol and tool, embodying a philosophy of order in the cosmos that could be harnessed for human movement.

What is striking across these cultures is the combination of precision and vulnerability. Stellar navigation depended on clear skies; a night of cloud cover could disorient a vessel. Seasonal changes altered which constellations were visible, demanding not just memorisation but adaptability. Polaris, so central for Mediterranean and Indian Ocean sailors, is invisible in the southern hemisphere, forcing Polynesians to develop alternative stellar frameworks. These limitations reveal that celestial navigation was never mechanical certainty but a balancing act between disciplined knowledge and unpredictable conditions. It demanded a form of intellectual resilience: the ability to cross-check stars with swells, winds, or bird movements when the sky failed. *The mind had to integrate multiple layers of observation into a coherent picture, a task more complex than following a single instrument.*

The philosophical implications are significant. To use stars as coordinates is to assume that nature is orderly, that celestial bodies move in patterns reliable enough to be trusted across lifetimes. This assumption is not trivial. It represents one of humanity's earliest commitments to rationality: the belief that by watching carefully, we can find structure in apparent chaos. For the Greeks, this conviction fuelled not only navigation but also cosmology; for Islamic scholars, it harmonised with theological views of divine order; for Polynesians, it confirmed that the heavens and Earth were connected through a knowable rhythm. The act of steering by stars was at once technical and philosophical, embodying a worldview in which the cosmos was intelligible, navigable, and open to human comprehension.

Modern comparisons sharpen the achievement. Today, GPS provides instantaneous positioning, requiring no perceptual skill from the user. Yet this reliance can create fragility: when systems fail, navigators are left disoriented. The recent revival of Polynesian voyaging, led by the *Hōkūle'a* canoe in the late twentieth century, demonstrates both the enduring value and the contemporary relevance of stellar navigation. Trained navigators successfully sailed from Hawaii to Tahiti without instruments, proving not only the feasibility of ancient methods but also their resilience. Their voyages were not exercises in nostalgia but reminders that navigation is a cognitive art as much as a technical one. In an age when tools externalise memory, the discipline of cultivating the sky as a coordinate system offers an alternative model: knowledge stored in the mind, refined through collective training, and tested against the real world.

The interplay of observation, abstraction, and transmission is key. Observation identified the stars and their movements; abstraction turned them into coordinate systems; transmission ensured that this knowledge endured across generations. Each culture expressed this interplay differently. Polynesians wove it into chants and rituals, Greeks into philosophical texts, Arabs into mathematical instruments. Yet the core achievement was the same: transforming the unreachable lights of the sky into a grid upon which human beings could plot their movement with extraordinary precision.

The lesson of stars as coordinates extends beyond seafaring. It illustrates how human beings turn perception into system, how disciplined attention can elevate raw experience into a reliable framework for action. It shows that technology need not always be material; sometimes the most powerful tools are intellectual structures carried in the mind. The starry sky, constant yet shifting, offered one of the first great challenges to human cognition. That it could be transformed into a navigational instrument is testimony to the ingenuity of ancient cultures—and a reminder that even today, surrounded by machines that compute positions in milliseconds, we remain heirs to those who first treated the cosmos as a map.

Waves and Wind as Clues

To sail thousands of miles of open ocean without a map or compass requires the cultivation of senses most modern navigators never develop. Pacific Islanders, who settled the remotest islands of the world centuries before European exploration, achieved this by learning to read the sea itself. The waves and winds were not background conditions but *textured signals*, carrying information about direction, proximity to land, and the invisible contours of the ocean. What looks to an untrained eye like an endless, undifferentiated surface was, to them, a patterned system of swells and gusts that could be read as surely as writing on a page.

Unlike stars, which are distant and stable, waves are dynamic, immediate, and localised. Their variability might seem to make them unreliable as navigational guides, yet it was precisely this variability that provided clues. In the vast Pacific, where winds blow steadily from certain directions for much of the year, the dominant swells reflect global patterns. At the same time, secondary swells created by local winds, storms, or the reflection of waves off island shores overlay these primary patterns. The skilled navigator could detect subtle shifts in rhythm, height, and angle, distinguishing the background pulse of oceanic trade wind swells from the cross-cutting interference of reflected waves. *The ocean was never featureless; it was structured, and those who knew how to read it carried a living chart beneath their canoes.*

Training began early. Children were taught to sit in canoes with their eyes closed, learning to feel the roll of the vessel and identify from which side the swell was striking. By adulthood, a navigator might discern three or more interacting swells simultaneously, constructing a mental model of intersecting lines across the sea. This model was portable and adaptive. At night, when stars were obscured by clouds, the navigator could rely on wave orientation to maintain course. Near land, where reflected swells bent back from island shores, these signals grew stronger, acting as a kind of invisible beacon. Navigators in the Marshall Islands went so far as to encode this knowledge into *stick charts*—woven frameworks of palm sticks with shells marking islands and curved lines denoting swell patterns. These charts were not taken on voyages but studied beforehand, serving as memory aids. The true chart was carried in the body, in the kinaesthetic memory of how a canoe moved across different wave structures.

The wind added a further layer of information. In equatorial and subtropical zones, trade winds blow consistently from east to west, providing a dependable orientation for voyagers. Yet changes in wind direction could signal seasonal shifts, approaching storms, or proximity to landmasses that disrupted airflow. Experienced navigators treated the wind not simply as propulsion but as context—

a companion element to the waves. If swells suggested one direction but winds contradicted them, careful judgment was required. Such discrepancies often meant that a storm system was altering conditions hundreds of kilometres away, reminding the sailor that navigation demanded not just spatial but also temporal awareness. Reading the sea was inseparable from anticipating its future states.

The intellectual achievement of this system lies in its integration of sensory input into a coherent framework. It was not a matter of instinct but of finely honed perception and abstraction. Navigators used micro-adjustments—the angle of a crest, the period between waves, the way a swell lifted or resisted the hull—to maintain bearings across enormous distances. Such calibration demanded a kind of mental endurance: hours, even days, of sustained attention without external instruments. *To navigate in this way was to cultivate a different relationship with knowledge, one where memory, sensation, and environment merged into a single system of orientation.*

Comparisons with European navigation highlight the distinctiveness of this method. When Portuguese and Spanish explorers first crossed the Atlantic and Pacific, they carried magnetic compasses, astrolabes, and eventually sextants. Their training emphasised measurement, calculation, and reference to written charts. By contrast, Pacific navigators worked with continuous feedback, not discrete measurements. Their method was not less rational but differently rational: it relied on embodied knowledge and dynamic modelling rather than written geometry. Where Europeans externalised knowledge into tools, Pacific Islanders internalised it into perception. This difference reveals a broader point: *engineering can be done with bodies as well as with instruments, and systems of knowledge need not be written to be precise.*

There were, of course, risks. Waves and winds are less fixed than stars. Storms could scatter swells, create false reflections, or obscure reliable patterns. In such cases, navigators needed to cross-check their readings against birds, clouds, or the smell of vegetation carried on the wind. Shipwrecks were not uncommon, and oral traditions preserve tales of voyages gone astray. Yet these stories were not only records of disaster; they were also educational, reinforcing the importance of vigilance and the limits of reliance on any single cue. The sea demanded humility. Its patterns were interpretable but never wholly predictable, and this uncertainty forced navigators to develop intellectual flexibility—a capacity to update their models in real time as conditions changed.

The modern revival of these techniques underscores their sophistication. In the late twentieth century, voyaging societies in Hawaii and Micronesia sought to recover traditional methods that had nearly vanished under colonial suppression. Navigators like Mau Piailug of Satawal trained new generations to feel swells and read winds without instruments. Their successful voyages across thousands of kilometres validated what anthropologists had long debated: that Pacific expansion was deliberate, not accidental. The *Hōkūle'a*, a double-hulled canoe launched in 1976, sailed from Hawaii to Tahiti guided solely by non-instrumental navigation, shocking sceptics and restoring pride in ancestral knowledge. These voyages demonstrated that ancient wave-reading was not only viable but competitive with modern tools, so long as the navigator had trained eyes, ears, and body.

The implications extend beyond seafaring. In an age dominated by externalised technology, the Pacific method reminds us of the capacity for humans to embody knowledge in perception itself. It challenges the dichotomy between intuition and calculation, showing that finely trained perception can achieve the rigour of science. The waves and winds are not random phenomena but lawful systems, accessible through experience. The intellectual leap lies in recognising that the sea, however fluid, is patterned—and that these patterns can be stabilised in memory and practice. To read the ocean is to accept its instability and still extract order, a lesson as relevant to modern complexity as to ancient voyaging.

To dismiss such knowledge as merely "traditional" is to miss its methodological depth. It required the construction of frameworks for distinguishing signal from noise, the training of attention to perceive subtle differences, and the transmission of this skill across generations without written texts. It demonstrates how a community can cultivate collective cognition—distributing expertise through apprenticeships, rituals, and oral instruction—so that the knowledge survives even when individuals perish. The canoe, then, was not simply a vessel; it was a mobile classroom, and the ocean itself a vast, unrolling manuscript written in swells and winds.

The genius of Pacific navigation lies in its capacity to make the sea speak. What to the uninitiated is chaos becomes structured, interpretable, and actionable. In that act of interpretation lies a profound intellectual achievement: turning the volatility of waves and winds into reliable guides, and doing so with nothing more than human perception honed by discipline. It was not technology in the modern sense, yet it was no less technological. It reminds us that knowledge is not confined to what can be inscribed on paper or encoded in machines. Sometimes it resides in the practised body, in the subtlety of senses trained to recognise patterns invisible to most. And in those patterns—rolling beneath the hull and streaming across the face—lay the keys to crossing the greatest ocean on Earth.

Mnemonic Navigation

To commit an ocean to memory is no small feat. Yet long before charts and compasses became standard instruments of seafaring, navigators developed ways of storing thousands of miles of spatial knowledge within the mind. Across Polynesia, Micronesia, and other seafaring cultures, routes were not written but sung, performed, and visualised. The sea became not only a physical expanse but a mental landscape structured by rhythm, story, and imagery. This was not improvisation but a rigorously maintained system of oral cartography—an architecture of memory sophisticated enough to rival any parchment map.

The challenge was immense. A voyage between islands could take weeks, with no landmarks for reassurance. To survive, navigators needed precise knowledge of bearings, expected distances, seasonal winds, star paths, and the swell patterns of different regions. In literate societies, such detail would be externalised onto paper. Here, however, memory itself had to serve as the archive. The solution was to harness the brain's natural affinity for narrative and rhythm. *Raw data are easily forgotten; woven into chant or story, they become enduring, repeatable, and transmissible across generations.*

The memory palace technique, known from classical Greece and Rome, has striking parallels in Pacific navigation. Orators in antiquity imagined architectural spaces, placing vivid mental images in sequence along corridors and rooms. Navigators, by contrast, constructed mental seascapes. An island might be associated with a mythical figure, a constellation with a remembered verse, a dangerous reef with a metaphor. In Micronesia, certain stars were grouped into "star lines," each line marking a directional axis that pointed toward particular islands. These were recited as part of a long sequence, such that knowing one's position was also a matter of knowing one's place in a story. The mnemonic scaffold was not abstract geometry but a narrative chain, binding astronomy, geography, and culture into a single whole.

Chants served as repositories of both information and authority. In Kiribati, navigators memorised extensive chants detailing the sequence of islands, the directions of ocean swells, and the qualities of different reefs. These chants were often performed in ritualised settings, reinforcing their seriousness and sacredness. To forget a line was not only a lapse of memory but a failure of responsibility, for lives depended on the accuracy of recall. *The human voice became the equivalent of a written chart, with rhythm and repetition functioning as safeguards against error.* Unlike paper maps, chants had to be embodied in the next generation; knowledge lived only if it was rehearsed.

Oral cartography was not limited to seafaring. Australian Aboriginal songlines provide an illuminating parallel. Songlines describe routes across vast landscapes, with verses encoding water sources, landmarks, and ancestral events. A singer could traverse hundreds of kilometres guided by the song's progression, each phrase tied to a physical feature. Though developed in a different environment, the logic was the same: narrative as navigation. Just as Polynesians sang the sea into coherence, Aboriginal custodians sang the land into orientation. These methods reveal a shared human ingenuity: the recognition that *memory is structured more effectively by story than by abstraction.*

The scepticism of early European explorers towards such methods is telling. Many dismissed the chants and stories as superstition, unable to see the layers of spatial coding embedded within them. Yet when rediscovered and tested in the twentieth century, these oral techniques proved astonishingly precise. In the case of Mau Piailug, the Micronesian master navigator who trained Hawaiians in the 1970s, his knowledge was not written but recited and demonstrated. He could steer across thousands of kilometres of open ocean, guided by nothing more than remembered sequences of stars, swells, and verses. The voyages of the canoe *Hōkūle'a* provided dramatic proof that oral cartography was not folklore but a practical science preserved in chant.

The philosophical depth of these systems deserves recognition. At one level, they are pragmatic tools for survival. At another, they reflect a profound insight: that knowledge is not only external but internal, and that memory, when structured correctly, can serve as infrastructure. Where modern societies trust devices and records, these navigators trusted their own cognitive architectures, built through ritual, community, and practice. *The chant was not a substitute for a map—it was a different kind of map, alive and dynamic, inseparable from the body that carried it.*

There were, of course, limitations. Memory is vulnerable to distortion, and oral traditions risk erosion if transmission falters. Disasters or misremembered verses could cost lives.

To guard against this, societies embedded navigation within cultural frameworks of prestige and responsibility. Only the most disciplined were entrusted with full navigational chants; training could take decades. This exclusivity acted as quality control, ensuring accuracy while binding knowledge to social identity. Navigation was not merely technical but moral—an office of trust within the community.

Modern cognitive science has shown why such methods worked. The brain excels at remembering sequences when they are tied to rhythm, imagery, and emotion. Mnemonic devices exploit neural predispositions, turning ephemeral data into durable patterns. What Pacific navigators discovered by necessity aligns with what psychologists now formalise in theories of memory encoding. Indeed, techniques such as the "method of loci" or modern memory competitions echo the principles of oral cartography. *The mnemonic navigator was, in effect, a human supercomputer—one that required no external storage because the storage was built into culture itself.*

Today, as GPS devices reduce navigation to passive following, there is renewed interest in these mnemonic systems. They remind us of what the mind is capable of when cultivated in the absence of technological crutches. They also raise broader questions: what kinds of knowledge do we neglect by over-relying on externalisation? If chants could hold oceans, what could our minds still hold if trained with similar rigour? To study mnemonic navigation is to encounter an alternative intellectual tradition, one that suggests that memory, far from being fragile, can be engineered into extraordinary resilience.

The endurance of these practices into the modern world shows their continuing relevance. Revivalist voyages, like those of the *Hōkūle'a*, not only validated ancestral methods but reasserted the dignity of oral knowledge in a technological age. They demonstrated that chants, memory palaces, and oral maps are not relics of a primitive past but evidence of human adaptability. The lesson is not that we should abandon charts and satellites, but that we should recognise the broader spectrum of human cognition. In the end, what the Pacific navigators achieved is a reminder that *knowledge is not limited to what can be written down; it can be lived, sung, and carried within the mind as reliably as ink upon paper.*

The Limits of Intuition

Confidence in one's sense of direction can be intoxicating, but on the open sea it has always been a dangerous luxury. Long before GPS satellites and digital cartography, the survival of entire crews depended upon the delicate interplay between memory, intuition, and external signs—stars, currents, clouds, and bird flight. Yet intuition, no matter how cultivated, was never infallible. For every successful voyage celebrated in song, there were wrecks that ended in silence. The fragility of "mental cartography" is revealed most vividly not in its triumphs but in its failures, where the human mind's bold attempt to master vast and shifting oceans encountered its own fallibility.

Consider the case of Portuguese navigators in the fifteenth century, who had extended their reach down the coast of Africa before daring to cross the Atlantic. Their ships were equipped with rudimentary charts, but much depended on dead reckoning—estimating position by speed, time, and heading. Overconfidence in such calculations often proved disastrous. Errors accumulated invisibly until they culminated in catastrophe: a landfall hundreds of miles off course, or worse, no land at all. Entire caravels were lost in the expanse between Europe and West Africa because navigators trusted the neatness of their reckonings rather than the messy reality of currents and winds. *The mind's map, elegant in conception, could be cruel in practice.*

The Pacific Islanders, whose feats of navigation are rightly celebrated, were not immune to failure either. While master navigators cultivated extraordinary precision, they too risked the inherent limits of memory and judgement. Chants misremembered, swell patterns misinterpreted, or skies obscured by weeks of storms could reduce even the most seasoned seafarer to uncertainty. Archaeological evidence suggests that not all settlement voyages ended in discovery; some canoes likely vanished without trace. Oral traditions often omit these failures, but silence itself hints at the price of human error. In cultures where oral cartography was the sole infrastructure, *the stakes of a lapse were absolute:*

there was no paper record to check, no device to reset, no second chance once the sea had swallowed the path.

Even in literate societies, intuition often carried dangerous weight. During the Age of Sail, European captains frequently relied on "sea sense"—the belief that experience alone could guide them through poorly charted waters. In 1707, this misplaced trust culminated in the Scilly naval disaster. A fleet of the Royal Navy, commanded by Sir Cloudesley Shovell, misjudged its position while returning from Gibraltar. Believing themselves safely west of Brittany, they in fact sailed straight into the Isles of Scilly. Four ships were wrecked, and nearly two thousand sailors drowned. The cause was not storm or battle but navigational uncertainty: the inability to determine longitude at sea. The tragedy was a turning point, leading the British government to establish the Longitude Prize in 1714. *The wreck exposed a painful truth—that no amount of intuition could substitute for an objective, verifiable measure of position.*

Philosophically, these failures illuminate the dual nature of mental cartography. On the one hand, it represents an extraordinary intellectual leap: the capacity to imagine unseen spaces, to convert environmental cues into an internalised geography. On the other hand, it reveals the cognitive biases that haunt human judgement. We are prone to overconfidence in our sense of direction, susceptible to anchoring on familiar patterns, and dangerously willing to ignore evidence that contradicts our expectations. A navigator might persist in a chosen heading not because the signs confirmed it but because turning back would mean admitting error. *The ocean punishes such biases mercilessly.*

This tension is not confined to history. Modern mountaineers, pilots, and even urban travellers exhibit similar overreliance on intuition. Studies in cognitive psychology show that people consistently misjudge distances and directions when deprived of external reference points. Our internal compass, though serviceable on familiar ground, is remarkably fragile in novel environments. The same neural machinery that allowed Polynesian navigators to hold entire archipelagos in memory is also capable of leading a hiker in circles within a dense forest. The difference lies not in biology but in training, discipline, and cultural reinforcement. Yet even the best training cannot erase uncertainty entirely; *it can only reduce the margin of error, never eliminate it.*

Shipwrecks and failures thus remind us that mental cartography, however ingenious, always balanced on a knife-edge between knowledge and assumption. To trust the stars, the swells, or the chant was also to accept that they could mislead or be misread. Cultures that thrived on oral navigation understood this implicitly, which is why they coupled knowledge with ritual, humility, and an acute sense of responsibility. Navigators were respected not only for skill but for their willingness to bear risk on behalf of the community. When voyages failed, the lesson was not simply technical—it was existential, underlining the limits of human mastery over nature.

The modern relevance of these historical failures is striking. Today's reliance on digital navigation tools offers extraordinary accuracy, but it also introduces new vulnerabilities. People who unthinkingly follow GPS directions have driven into rivers, onto runways, or into deserts because they outsourced judgement entirely to technology. These are, in their own way, contemporary shipwrecks—failures not of intuition but of its absence. The pendulum has swung from overconfidence in the mind's map to blind faith in the machine's. Both extremes reveal the same underlying danger: an abdication of critical balance between internal judgement and external verification.

The lesson is not to dismiss mental cartography but to understand its boundaries. Human intuition is capable of astonishing feats, as demonstrated by the navigators who crossed oceans without instruments. Yet its failures remind us that *intuition without checks is perilous*. The intellectual leap of mapping the unseen requires both humility and vigilance. The mind can chart seas, but the sea will always remind us of its power to resist certainty.

Chapter 3
Raising Stone Without Steel

To stand before an Egyptian pyramid or a Mesopotamian ziggurat is to confront a paradox. How did societies lacking cranes, steel tools, or industrial cement manage feats that continue to astonish modern engineers? The achievement was not simply in lifting stone, but in orchestrating matter, mind, and meaning into enduring form. Monumental construction in antiquity combined physics, imagination, and social ingenuity in ways that were both practical and profound.

Mechanical principle offers one of the clearest windows into this ingenuity. Long before Archimedes claimed he could move the world with a lever, Egyptian engineers were exploiting balance and leverage to render colossal tasks manageable. Rolling a multi-tonne stone across desert sands demanded more than brute force; it required understanding that friction could be tamed with water, that inclined planes could multiply human strength over distance. Wall reliefs depicting workers pouring water before sledges are not mere illustrations—they are evidence of intuitive experimentation, an applied mechanics centuries before Newton. These techniques reveal that intelligence is not measured solely in abstract theory but also in the capacity to translate physical principles into practical solutions.

Yet the technical mastery of the ancients was inseparable from symbolic purpose. Sacred geometry underpinned construction, providing a scaffolding as much philosophical as structural. Temples and pyramids were aligned with celestial bodies and cardinal points, embedding religious cosmology into stone. Mesopotamian ziggurats rose as layered reflections of the universe itself. In these cultures, engineering and theology were intertwined: mathematics conveyed meaning, and geometry became a medium of metaphysics. Walking through such spaces was to enter a world in which balance and harmony were both spiritual and structural. The stones were not merely moved—they were ordered, imbued with symbolic coherence.

Labour, too, was a form of technology. It is easy to reduce construction to levers and ramps, but the organisation of vast workforces was itself an innovation. Coordinating thousands of people required hierarchies, rituals, and shared purpose, aligning human effort both physically and psychologically. In the pyramid, administration was as vital as engineering; the workforce functioned as a living machine, capable of feats beyond any individual. The true ingenuity lay in transforming social structures into extensions of architectural technique.

Centuries later, the Romans would add chemistry to this repertoire. Their concrete—volcanic ash, lime, and rubble—has endured millennia, often outlasting modern counterparts. Unlike contemporary cement, which can crumble within decades, Roman concrete grows stronger with time, thanks to chemical interactions with seawater and air. Achieving such durability without analytical chemistry demonstrates another form of knowledge: empirical intuition honed over generations. Builders observed, experimented, and preserved what worked, embedding science within craft. To walk beneath a Roman aqueduct or the Pantheon is to witness chemistry written not in equations, but in enduring stone.

Viewed together—leverage and balance, sacred geometry, labour as technology, and empirical chemistry—these achievements reveal a holistic approach to problem-solving. Ancient monuments were not merely scaled-up buildings; they embodied the integration of physics, philosophy, and social coordination. The intellectual architecture that raised the stones is as remarkable as the stones themselves. Technology, in this context, is never merely material—it is cultural, cognitive, and ethical, shaped by imagination and values.

For the contemporary reader, these lessons resonate beyond historical curiosity. The durability of Roman concrete challenges assumptions about progress, suggesting that ancient intuition can surpass modern science in longevity. Egyptian labour organisation offers insights into managing complexity in large-scale projects. Sacred geometry prompts reflection on whether today's architecture, driven by speed and economy, has lost the capacity to convey deeper meaning. Leverage and balance remind us that fundamental principles, carefully observed, underpin even the most advanced technologies. Studying how the ancients raised stone without steel teaches how human ingenuity, confronted with constraint, generates insight that transcends those limits.

This chapter, therefore, is not an indulgence in wonder. It is an exploration of ancient construction as an intellectual enterprise, one that sharpens understanding of problem-solving, design, and durability. By examining how Egyptians, Mesopotamians, and Romans mastered stone, we uncover lessons about the interplay of mind, matter, and meaning. The monuments endure, but more crucially, the methods by which they were raised illuminate a deeper continuity: human ingenuity reaches its peak when mechanical insight, symbolic order, social coordination, and empirical craft converge into a unified vision. In that convergence lies the architecture of thought itself, set in stone for the ages.

Leverage and Balance: How Egyptians and Mesopotamians Applied Mechanical Principles to Move Colossal Stones

The heaviest stone blocks ever quarried and set into place were not the work of modern cranes or industrial machinery but of ancient minds that turned necessity into invention. In the Egyptian temple complex at Baalbek, blocks weighing upwards of 800 tonnes were cut, transported, and positioned with an accuracy that astonishes engineers even today. The pyramids, composed of millions of blocks, each weighing several tonnes, rise with such symmetry and precision that their alignment with cardinal directions remains one of history's great feats of applied knowledge. The paradox is clear: without iron tools, pulleys, or wheels suited to such burdens, how did early civilisations marshal stones of unimaginable weight into enduring forms? The answer lies not in fantastical theories but in the disciplined, intuitive grasp of basic mechanical principles: leverage, balance, and the reduction of friction.

The Egyptian approach to moving stone reveals a society that combined acute observation of natural forces with the ability to scale simple methods to extraordinary dimensions. The sledge, a seemingly humble device, exemplifies this. By dragging stones across sledges lubricated with water, workers reduced friction dramatically—a fact demonstrated in modern experiments that show dampened sand can cut resistance by as much as half. What reads in textbooks as "reducing the coefficient of friction" was for the Egyptians an embodied understanding, refined through trial and error, that weight could be made to yield to human force if the right medium was applied. The insight was not theoretical but practical, drawn from the physical experience of sand, water, and timber under load. *This was science expressed in muscle memory rather than equations.*

Equally striking was the use of ramps, whose forms—straight, zigzagging, or encircling—remain subjects of scholarly debate. Whatever their configuration, ramps translated the vertical challenge of lifting into the horizontal domain of dragging. The principle is deceptively simple: the inclined plane distributes effort over distance, allowing a team of workers to move weights otherwise immovable. Here again lies an elegant truth. The pyramid was not raised by lifting blocks into the air but by lengthening the path upon which they could be shifted, transforming steepness into gradual ascent. In this we see a civilisation not overawed by scale but capable of reimagining scale in terms of proportion and geometry. By extending the path, they extended their power.

Mesopotamia, though lacking the monumental stone quarries of Egypt, demonstrated comparable ingenuity in the movement and placement of heavy loads. Ziggurats, rising in staged platforms, required not only brick by the million but also immense foundation stones and structural supports. Reliefs from the reign of Ashurnasirpal II depict teams of workers hauling colossal statues and stones, employing ropes, levers, and sledges in coordinated effort. The lever, one of the simplest machines in mechanical science, was indispensable. By inserting wooden beams beneath stones and using fulcrums, Mesopotamian builders could shift massive objects incrementally, each small movement amplified by patience and precision. What seems laborious to modern eyes was in fact an efficient strategy, grounded in the principle that force multiplied by distance allows even modest inputs to achieve vast outcomes.

There is a philosophical dimension to these methods that deserves attention. The act of moving stone was not solely a technical puzzle but an intellectual confrontation with weight, scale, and permanence. To build at such magnitude required confidence in principles that could not be seen but only felt. A lever works because of a balance of moments; a ramp succeeds because resistance is redistributed over space. These are abstractions, yet they were grasped by builders who had no symbolic language for torque or vectors. *Theirs was an intuition of physics before physics, an apprenticeship to nature's laws carried out through the body.* This suggests that human cognition is capable of profound insight without formal systems of notation, provided that observation and necessity press hard enough.

Modern experiments have repeatedly confirmed the plausibility of such techniques. Archaeologists and engineers have reconstructed sledges, ramps, and lever systems to demonstrate that teams of workers could indeed achieve feats once deemed impossible. Mark Lehner's work at Giza has shown that cooperative groups, rotating shifts, and coordinated hauling could move blocks up ramps with surprising efficiency. Japanese researchers have recreated stone-pulling exercises with wetted sand, confirming the reduction of friction described in Egyptian reliefs. Such experiments not only validate ancient practices but also expose the fallacy of assuming that only advanced machinery can account for monumental scale. The persistence of fringe theories invoking extraterrestrial intervention or lost technologies reveals more about modern scepticism than ancient capability.

Yet to reduce these methods to mechanics alone risks overlooking the broader innovation: the ability to scale knowledge socially. A lever or ramp functions in the hands of a single individual, but to apply it to a project spanning decades and involving tens of thousands of workers requires coordination, discipline, and foresight. The Egyptians and Mesopotamians transformed simple machines into civilisation-scale solutions by embedding them into systems of labour and ritual. Workers chanting in unison while hauling ropes were not merely engaging in song but synchronising their force, aligning rhythm with exertion. Overseers calculated the alignment of ramps, the angle of incline, and the staging of shifts with administrative precision. Thus, leverage and balance were not only physical principles but social ones: balance of effort across time, leverage of human numbers into cumulative achievement.

The modern relevance of these insights is more than symbolic. In an age where engineering is often equated with computational modelling and digital precision, the ancients remind us that *the first principles of mechanics remain the bedrock of innovation.* Bridges, skyscrapers, and spacecraft all rely upon the same underlying truths that moved stones at Giza and Babylon. Engineers today calculate load distribution and torque with mathematical formalism, but the principles are unchanged. What differs is the medium of understanding: equations for us, embodied practice for them. To recognise this continuity is to acknowledge that progress is not the discovery of new laws but the refinement of how we articulate and apply them.

There is also a humbling lesson in durability. The pyramids remain upright not simply because of their mass but because the methods of construction aligned with the realities of material and force. They were not designed to defy physics but to cooperate with it. In contrast, many modern constructions, built with speed and efficiency in mind, falter within decades. Leverage and balance, as understood by the ancients, extended beyond the practicalities of stone-moving into a philosophy of endurance: build in harmony with natural law, and your work will outlast you.

To study how Egyptians and Mesopotamians moved stone is therefore to study how humans transform constraint into creativity. They did not wait for tools they lacked; they expanded the capacity of the tools they had. They treated force not as an adversary to be overcome but as a partner to be rechannelled. In the rhythm of hauling ropes, the slope of a ramp, or the careful placement of a lever beneath a stone, we glimpse a relationship between human will and natural law that is at once pragmatic and profound. What was raised was not only stone but an enduring testament to the power of mechanical insight harnessed at civilisational scale.

Sacred Geometry: How Religious Cosmologies Shaped Architectural Proportions into Physical Philosophies

The stones of ancient temples were not simply stacked to shelter ritual; they were laid to embody truths about the cosmos. For Egyptians, Mesopotamians, Greeks, and later Romans, architecture was not merely a matter of engineering but a medium through which metaphysical order was made tangible. The striking regularity of their monuments—the square courtyards, symmetrical facades, carefully calculated ratios—was neither accidental nor aesthetic indulgence. It was an attempt to align human space with the structure of the universe, turning sacred geometry into an organising principle of both thought and stone. *To walk into these temples was to step inside a worldview, measured in cubits and chords rather than words.*

Egypt provides one of the clearest illustrations of how geometry and cosmology fused. The orientation of the Great Pyramid at Giza is aligned with uncanny accuracy to the cardinal directions, suggesting not only mathematical ability but a profound concern with situating human structures within celestial order. The pyramid form itself carried multiple symbolic resonances: its base represented the earth, its apex the point of contact with the heavens, while its sides captured the rays of the sun descending to earth. This was not geometry for its own sake, but geometry as a conduit of meaning—each proportion and angle a philosophical statement expressed in stone. Egyptian temples likewise reflected cosmic dualities. The sanctum was often conceived as the primeval mound emerging from the waters of chaos, its proportions echoing myths of creation. Columns designed like lotus or papyrus stems became metaphors in limestone for the order of nature. *Every measure was both structural necessity and cosmological symbol, inseparably bound.*

Mesopotamian ziggurats, though starkly different in form, reveal a parallel preoccupation with proportion as philosophy. Their tiered structures rose like artificial mountains, linking earth to heaven step by step. Each level represented a layer of the cosmos—the subterranean waters, the earthly plane, the heavens above—rendering theology into architecture. The measurements of these tiers were not arbitrary: inscriptions suggest priests prescribed ratios that mirrored cosmic order, ensuring the building itself became a kind of calendar and cosmogram. Approaching a ziggurat was therefore a passage through theology, as if the brickwork itself instructed the devotee in the structure of the universe.

The Greeks, inheritors and refiners of this impulse, articulated sacred geometry with extraordinary precision. The Parthenon embodies ratios—most famously the golden ratio and Pythagorean harmonics—that suggest a deliberate attempt to bind architecture to universal laws of harmony. Columns are not simply vertical supports but carefully proportioned to taper subtly, correcting optical illusions while simultaneously manifesting balance and rhythm. Greek architects spoke of symmetry and harmony not in casual terms but as principles that reflected the same order thought to govern the cosmos. Pythagoras had taught that numbers were the essence of reality, and architects responded by embedding numerical harmonies in stone. The Parthenon, with its fusion of art, religion, and mathematics, becomes less a civic building than a philosophy of order given physical form.

The Romans, pragmatic by temperament, nonetheless extended this tradition in their own way. Vitruvius, in *De Architectura*, codified the principle that buildings must embody *firmitas* (strength), *utilitas* (function), and *venustas* (beauty), with beauty arising from correct proportion. His insistence that architecture be modelled on the proportions of the human body reflects a cosmological idea: man as microcosm of the universe. The circle and square—shapes recurring in Roman temples and forums—were not decorative preferences but emblems of perfection and universality. When Brunelleschi and later Renaissance architects rediscovered Vitruvius, they did not merely revive Roman engineering but reignited this metaphysical conviction that geometry was a language of truth.

What unites these civilisations is the conviction that geometry was more than a technical instrument; it was a bridge between the human and the divine. Proportion was a moral as well as aesthetic matter. A temple badly aligned was not simply an architectural failure but a philosophical transgression, a refusal to harmonise with cosmic law. This imbued the role of the architect with unusual authority. He was not merely a craftsman but an interpreter of the universe, tasked with ensuring that human structures resonated with eternal order. The fact that these structures endure millennia later is testimony not only to their engineering but to the depth of the principles upon which they were founded.

Modern architecture, for much of the twentieth century, turned sharply away from these convictions, embracing functionality stripped of symbolism. Le Corbusier, however, attempted to revive a vestige of sacred geometry through his "Modulor" system, an anthropometric scale based on the golden ratio. His aim was to reconcile human proportions with architectural design, echoing Vitruvius while casting it in a modern idiom. The fact that his system provoked both admiration and critique highlights a tension that has persisted since antiquity: whether geometry in architecture is merely instrumental or inherently metaphysical. The glass towers of modern cities, built to efficiency rather than cosmology, seem to deny any sacred dimension. Yet their uniform grids, often unintentionally, replicate the same mathematical obsessions with symmetry and proportion, even if shorn of overt symbolism.

There is also renewed interest in sacred geometry among contemporary architects who seek to reconnect with ideas of harmony and sustainability. Islamic architecture, for example, preserves geometric patterns that symbolise infinity and unity, embedding spiritual cosmology in visual form. The Alhambra's intricate tiling systems reflect not only aesthetic mastery but theological concepts of divine order and infinity. Today, designers inspired by these traditions explore fractal geometries and biomimetic forms, recognising that *geometry is also the language of nature.* What the ancients expressed through myth and ratio, modern architects can test through mathematics and material science, often arriving at similar insights: that proportion creates both beauty and endurance.

The study of sacred geometry forces us to reconsider the boundaries between science, religion, and art. For the ancients, these divisions did not exist; geometry was simultaneously a practical tool, a spiritual language, and a philosophical inquiry. To walk through Karnak or the Parthenon was not to enter a neutral building but to participate in an argument about the nature of reality, expressed not through words but through ratios of stone. This is why sacred architecture retains its impact even on secular visitors. One does not need to believe in Egyptian creation myths or Greek cosmology to feel the weight of order and harmony in these spaces. *The geometry itself communicates across belief systems, whispering of a universal order that transcends cultural particularity.*

Critics may object that not every proportion or alignment was intentional, and caution is warranted against over-interpretation. Archaeologists debate whether claims about the golden ratio in the Parthenon are retrospective impositions rather than deliberate design. Some argue that what we call "sacred geometry" was often simply practical craftsmanship, honed by habit rather than metaphysics. Yet even if not every measurement was symbolic, the cumulative pattern across civilisations points to a consistent entanglement of geometry with cosmology. To deny this entirely would be to understate the intellectual ambition that these societies poured into their most significant works.

Sacred geometry thus reveals a dimension of architecture that is easy to overlook in modern discourse: the power of proportion not only to please the eye but to shape consciousness. The ancients understood that to inhabit space was to be formed by it. A temple aligned to the stars or measured to the rhythm of numbers trained its visitors to see order where chaos might otherwise intrude. It reminded them that the human world was nested within a larger cosmos. In this sense, temples were not simply places of ritual but instruments of orientation, both physical and metaphysical. They taught, silently yet persuasively, that to live well was to live in proportion.

To trace the story of sacred geometry is therefore to encounter a history of human beings striving to embody in stone the patterns they believed governed existence itself. The lines, angles, and ratios were not abstractions; they were convictions that shaped the very spaces where life, worship, and politics unfolded. By turning geometry into architecture, ancient builders created a medium that spoke to both body and spirit, ensuring that the stones themselves carried forward philosophies long after their original priests and philosophers were gone.

Labour as Technology: Human Organisation as the Engine of Ancient Construction

The towering monuments of antiquity—whether the pyramids of Egypt, the ziggurats of Mesopotamia, or the amphitheatres of Rome—are often admired as feats of engineering or craftsmanship. Yet the most significant innovation that enabled these structures was not a tool of stone or metal, but the deliberate organisation of people. Before steel cranes or hydraulic presses, civilisations learned to transform human effort into a kind of living machine. *Labour itself became technology—structured, disciplined, and ritualised into an instrument of power and creation.*

Egyptian pyramids remain the archetypal example. For centuries, the prevailing image was of a vast slave force driven under the lash. Archaeological discoveries, however, have replaced this narrative with a more sophisticated reality. Excavations near Giza reveal purpose-built workers' villages, bakeries, breweries, and medical facilities, suggesting that the workforce was not an enslaved underclass but a rotating body of skilled labourers, perhaps conscripted seasonally during the Nile's inundation. Organisation here was key: tens of thousands of individuals had to be fed, housed, coordinated, and kept motivated for decades. The stones themselves, weighing several tonnes each, could not have been raised by brute force alone. It was the hierarchies of oversight, the division of labour into teams, and the ritual dimension of collective effort that made construction possible. In effect, the pyramid was not just an artefact of stone, but the by-product of an entire society's capacity to organise itself.

Mesopotamian ziggurats illustrate the same principle in a different register. The mud-brick core of these artificial mountains was not technically complex, but the sheer scale of their construction demanded extraordinary administrative coordination. Clay had to be quarried, shaped, dried, and transported in astonishing quantities. Bureaucracies of scribes recorded rations for workers, tallied resources, and enforced quotas. The very invention of cuneiform writing, some scholars argue, was stimulated by the need to account for labour and materials in such large projects. Here, organisation was not merely applied to labour—it generated new technologies of administration that would transform the entire civilisation. *The building of temples became the building of systems.*

Ritual played an equally crucial role in transforming labour into technology. Human bodies can be driven by coercion, but endurance requires meaning. Egyptian festivals and Mesopotamian rites infused construction with sacred purpose, casting the worker not merely as a labourer but as a participant in cosmic renewal. In some cultures, songs, chants, and coordinated rhythms were used to synchronise motion, allowing hundreds of individuals to pull in unison as if they were parts of a single machine. These rituals had a dual effect: they reduced physical strain through coordination while reinforcing the psychological commitment of the labour force. In this way, ritual was both a spiritual technology and a practical mechanism of efficiency.

Hierarchy was another indispensable tool. At Giza, graffiti left by workers reveal the presence of organised crews with names such as "The Friends of Khufu." Each group was divided into smaller units, supervised by foremen who reported to higher officials, culminating in the master architect and ultimately the pharaoh. This fractal pattern of organisation resembles a chain of gears, where small, repeatable units of labour meshed into a vast, coherent force. Such structures of oversight were not incidental; they were themselves inventions, as critical as ramps or rollers. Indeed, one can argue that the true legacy of pyramid-building was not architectural but administrative: the capacity to marshal massed populations for coordinated, long-term projects.

The Romans took this principle further, integrating labour organisation into the very fabric of their empire. Roads, aqueducts, and amphitheatres were not isolated projects but parts of a continuous system of construction sustained by legions, enslaved populations, and paid artisans. Military discipline provided a template: soldiers could march, dig, and build with precision, transforming themselves from warriors into engineers as required. The Roman camp, laid out with mathematical regularity wherever the army paused, was itself an expression of labour as technology—a mobile architecture built daily through collective discipline. Roman monuments, from the Colosseum to the aqueducts, must therefore be seen not only as works of stone but as crystallisations of military-administrative labour systems.

What is striking is how these ancient experiments in organisation anticipate principles later formalised in management science. Division of labour, introduced to the modern canon by Adam Smith, was already implicit in pyramid construction. Bureaucratic record-keeping, later analysed by Max Weber, has its prototypes in Mesopotamian temple economies. Even Frederick Taylor's "scientific management," with its insistence on efficiency and coordination, echoes the ritualised rhythms of ancient workforces pulling stone blocks in synchrony. *Long before machines multiplied human power, organisation multiplied human possibility.*

There are tensions, however, that must be acknowledged. To describe labour as technology risks overlooking the human cost. The same organisation that produced monumental beauty also entrenched hierarchies of power, often coercive in nature. Workers at Giza may have been fed and housed, but their lives were still commandeered for the glorification of rulers. Roman spectacles, funded by colossal building projects, were made possible by enslaved populations whose exploitation was integral to the system. The brilliance of organisation is inseparable from its darker legacy: the capacity to harness human beings not only as citizens but as instruments.

Modern parallels make this legacy uncomfortably clear. The skyscraper rising from a contemporary skyline depends less on novel tools than on supply chains, managerial hierarchies, and coordinated labour forces distributed across the globe. When a smartphone is manufactured, it is not the robot on the assembly line that constitutes the deeper innovation, but the planetary system of logistics and human organisation that makes its production possible. The principle is the same as at Giza or Babylon: *labour, when structured and synchronised, becomes a form of technology in its own right.*

Yet there is also an instructive difference. Ancient systems infused labour with shared meaning, however coercive or ritualised, whereas modern systems often fragment and alienate. A worker hauling stone for a pyramid could believe he was helping maintain cosmic order. The modern labourer, part of a global supply chain producing components for an unseen device, rarely enjoys the same coherence of purpose. The ancients remind us that labour organisation, at its best, did more than build; it also bound people into a shared project that transcended the individual.

The history of monumental construction thus forces us to expand our definition of technology. Innovation does not lie only in tools of metal and stone but in the structuring of human effort. The ramp, the lever, the pulley—these were crucial, but without the living system of labour organisation they would have remained inert. What animated them was the collective body, disciplined through hierarchy, ritual, and meaning. The ancients did not merely build monuments; they built the organisational forms that would echo through empires, economies, and eventually the globalised systems of the present.

To look at the pyramids, ziggurats, or aqueducts today is to witness not simply feats of engineering but embodiments of a deeper principle: that human beings, when organised with clarity and purpose, can become the most powerful technology of all.

Concrete Before Cement: The Roman Alchemy of Stone and Longevity

Walk along the harbour of Cosa on Italy's Tyrrhenian coast and you tread upon concrete poured more than two millennia ago. Waves have battered it, salt has seeped into it, and yet it endures with a solidity that mocks many modern breakwaters crumbling after a few decades. The Pantheon in Rome, with the world's largest unreinforced concrete dome, has stood since the reign of Hadrian. The aqueducts that snake across Europe still carry water. These are not curiosities of preservation but evidence of a material that behaved differently, aged differently, and embodied an entirely distinct philosophy of building. Roman concrete—*opus caementicium*—was not simply a precursor to modern Portland cement. It was a fundamentally different material, devised without scientific laboratories, whose endurance continues to puzzle engineers. The Romans did not fully understand why their mixtures worked; yet their empirical intuition achieved a chemistry that modern science is only now beginning to decode.

Modern concrete is designed for speed and standardisation. Portland cement, the dominant binding agent since the nineteenth century, hardens quickly and allows for mass production. Yet its lifespan is notoriously short. Bridges, roads, and apartment blocks demand repairs within decades, their steel reinforcements corroding as water infiltrates. By contrast, Roman concrete appears almost alive in its durability. Studies have shown that when cracks form in Roman concrete exposed to seawater, they do not spread uncontrollably; instead, the material gradually heals itself, as if possessing a form of resilience coded into its mineral structure. *The secret was not precision science but a shrewd engagement with the materials at hand, guided by observation and long memory.*

At the heart of Roman mixtures was volcanic ash—*pozzolana*—sourced particularly from the Bay of Naples. When combined with lime and water, and often mixed with rubble of tuff or brick, the ash triggered a pozzolanic reaction that produced calcium–aluminium–silicate–hydrate, a compound that resists chemical decay. This reaction is slow, but it yields remarkable stability over centuries. Modern concrete, reliant on the rapid hydration of Portland cement, sacrifices such long-term robustness for immediate strength. The Romans were playing a different temporal game: they built not for decades but for centuries, and their materials were chosen accordingly.

The interaction with seawater is especially revealing. Contemporary engineers fear salt as the nemesis of concrete, corroding reinforcements and expanding cracks. Roman maritime concrete, however, harnessed seawater as an ally. Mineral analyses reveal that over time, contact with seawater initiated further crystallisation of phillipsite and aluminous tobermorite, rare minerals that grew within the concrete matrix. These crystalline structures filled gaps and effectively locked the material together more tightly with age. Where modern concrete weakens, Roman concrete strengthens. It is a reminder that durability is not merely a matter of resisting nature but of learning how to collaborate with it.

This raises a philosophical point: the Romans treated materials not as inert matter to be dominated, but as partners in a process. Their builders operated through trial, observation, and tradition, refining recipes over generations. Vitruvius, the Roman architect and engineer, gave recipes for mixtures in *De Architectura*, but these were not universal formulae; they varied by region, adapted to local resources. In this sense, Roman concrete was not a product but a practice, inseparable from the ecological and geological contexts in which it was made. *Chemistry by intuition meant chemistry by relationship—an embodied knowledge of how materials behaved in different conditions, honed by collective memory rather than laboratory instruments.*

The contrast with modern methods is instructive. Industrial concrete aspires to universality: a formula that works everywhere, scalable across continents. Yet in stripping away the particularities of local geology, it often sacrifices resilience. Roman builders did the opposite. They leaned into the specificity of place, trusting that the ash of Campania or the lime of the Tiber valley would yield distinct strengths. Their monuments were therefore deeply local, even as they served imperial purposes. The durability of the Pantheon or the harbours of Ostia is not an accident of genius but the outcome of an attentive dialogue with nature's resources.

One might object that Roman concrete was limited. It could not be poured in high-strength forms like today's reinforced concrete. It was weaker in tensile capacity, which is why Roman architecture relied on arches and vaults rather than beams. But this supposed limitation became an engine of creativity. Arches and domes distributed weight efficiently, producing structures of grace and monumental scale. The Pantheon's dome, lighter at the crown due to gradual changes in aggregate material, demonstrates how material and form were reconciled in a synergy rarely seen today. What we interpret as stylistic innovation was in fact structural adaptation to the material's properties.

The endurance of Roman concrete poses uncomfortable questions for modernity. Why, with advanced chemistry and global industries, do we produce a material that fails so quickly? One answer lies in priorities. Roman builders valued longevity and continuity. Their cities and ports were imagined as enduring fixtures of empire. Our economies, by contrast, are driven by cycles of growth and replacement. A building designed to last only fifty years is not a failure but a profitable cycle of construction and demolition. Planned obsolescence is embedded not only in consumer electronics but in concrete itself. Roman concrete reminds us that durability is not a technical impossibility but a cultural choice.

Recent research has attempted to recover Roman methods. Laboratories at MIT and elsewhere have examined the microstructures of ancient samples, identifying lime clasts that may have served as reservoirs of reactive material, allowing self-healing when cracks formed. Experimental reconstructions using volcanic ash and lime mixtures have produced concretes that indeed gain strength over time. Yet scaling these methods for contemporary construction faces resistance: the recipes are slower, less uniform, and incompatible with the pace of modern industry. The Romans were patient; their walls took months or years to cure fully. Today's construction timelines leave no room for such temporality.

There is, however, a growing awareness that the Roman approach holds relevance beyond archaeology. As climate change accelerates, the environmental cost of Portland cement has come under scrutiny. Cement production is one of the largest industrial sources of carbon emissions, releasing vast amounts of CO_2 through the calcination of limestone. Roman methods, which used less lime and more natural pozzolans, were less carbon-intensive and produced materials that needed replacement far less often. In ecological terms, durability is sustainability. The ruins of Rome suggest that an ancient intuition may offer solutions for the most urgent engineering challenges of the present.

Consider also the symbolic dimension. Roman concrete was not hidden behind facades but celebrated. The *opus caementicium* core of aqueducts and amphitheatres was often visible, displaying its rugged texture. It was not meant to simulate stone but to manifest a new material identity. In this sense, Roman concrete was both pragmatic and philosophical: a declaration that humanity could reconstitute nature itself, binding stone, ash, and water into a new geological substance. Modern concrete, smoothed and homogenised, often conceals its presence, aspiring to vanish behind glass and steel. The Roman material, by contrast, insists on its presence—ageing, cracking, and healing in ways that remind us of time's passage.

If there is a lesson to be drawn, it is that *intuition can be a form of science, provided it is iterative, collective, and attentive to context.* Roman engineers lacked theories of crystallography or mineralogy, but they possessed something equally potent: the ability to watch how materials behaved, to remember across generations, and to encode that knowledge into practice. Their concrete was not a triumph of sudden discovery but of sustained attention. That, perhaps, is why it endures.

To walk beneath the Pantheon's dome or touch the harbour walls of Cosa is to feel a quiet rebuke. Our age, saturated with technologies, often struggles to produce materials that survive a single human lifetime. The Romans, with fewer tools but deeper patience, produced structures that have outlasted empires. Their legacy is not only material but conceptual: a reminder that true innovation does not always come from precision instruments or rapid invention, but from the capacity to observe, to adapt, and to build for futures we may never see.

Chapter 4
Feeding Millions Without Chemistry

The greatest technological achievement of the ancient world was neither the pyramid nor the aqueduct, nor the bronze weapon or the wheel. It was the capacity to feed millions without synthetic fertilisers, chemical pesticides, or industrialised processes. Entire civilisations depended on the art of coaxing sustenance from earth and sky, a feat accomplished not in factories but through disciplined thought, observation, and cultural memory. Securing reliable food demanded more than sheer labour; it required imagination, measurement, foresight, and organisation on a scale at once scientific and philosophical. Agriculture was not a passive rhythm of sowing and reaping but a sustained intellectual enterprise. Domestication of plants and animals, management of rivers, tracking of celestial cycles, and preservation of surplus formed a suite of innovations as radical in their day as artificial intelligence or genetic engineering feels to us now.

Domestication, often presented as a natural process, was in fact a deliberate invention. Early farmers in the Fertile Crescent did not simply stumble upon wheat as we know it; they transformed wild grasses with fragile stalks and meagre kernels into stable, high-yield crops through selective replanting. This was not luck but centuries of experimentation, choosing seeds with larger grains, non-shattering ears, and predictable maturation. Animals were similarly shaped: cattle that tolerated confinement, sheep producing manageable fleece, dogs attentive to human command. Each choice represented a hypothesis tested across seasons, even generations. Domestication was the first long-term science project, enacted not in laboratories but in fields and pastures. Waiting decades for traits to stabilise required an intuitive understanding of inheritance and variation, a prelude to genetics millennia before the concept would be formalised.

Yet controlling life in the soil was only the beginning. Water, unpredictable and unforgiving, had to be disciplined. Mesopotamian farmers carved canals to channel the Tigris and Euphrates, turning floodplains into arable land. Egyptians timed planting to the rhythms of the Nile, whose inundation they both revered and feared. In the Andes, the Inca sculpted terraces along steep mountainsides, creating microclimates that maximised productivity. These interventions were not mere feats of engineering; they were acts of reasoning, demanding a deep understanding of hydrology, soil, and slope, coupled with governance systems to maintain canals and dykes. Irrigation, in other words, was as social as hydraulic, requiring coordination, negotiation, and rules. Unlike modern systems that often degrade soil or exhaust aquifers, ancient water management reflected a logic of sustainability: knowing when to release water, when to withhold it, and when to allow floods to replenish the land.

If irrigation demonstrates control over the horizontal plane, calendars reveal mastery of the vertical: time itself. To plant without regard for season is to gamble blindly. Ancient astronomers monitored the sun, moon, and stars not for curiosity alone but to determine the precise moments for sowing and harvesting. The Egyptians noted the heliacal rising of Sirius, signalling the Nile's flood; the Maya tracked Venus to guide maize cultivation; across Eurasia, festivals aligned with solstices and equinoxes. In these practices, cosmology and ecology were inseparable. Reading the heavens meant reading the future of the fields. Calendars were not passive records but active tools for synchronising labour with natural cycles. Where satellites and meteorological models now forecast rainfall, ancient farmers relied on stars and shadows, their accuracy measured in harvest yields rather than datasets.

Abundance, however, was useless without storage. The invention of granaries was, in many ways, the invention of civilisation itself. Grain that spoiled or was devoured could not sustain cities or armies. Societies developed elevated silos, clay-lined pits, and sealed structures to protect food. Preservation extended beyond grain: fish was salted in Phoenician ports, fruits sun-dried in Mesopotamia, and fermentation stabilised nutrients. What emerges from these practices is a sophisticated form of engineering: mastery over time. Storage allowed food to outlast seasons, smoothing the variability of harvests and enabling surpluses that freed some to become artisans, bureaucrats, or soldiers. The granary was not a mere warehouse but a temporal device, shifting calories from moments of plenty into periods of scarcity. Preservation techniques, in this sense, were as critical as irrigation canals or ploughs, transforming agriculture from crisis management into a dependable system.

Viewed together—domestication, irrigation, calendars, storage—ancient agriculture appears as an integrated framework of knowledge rather than simple labour. Each innovation was experimental, empirical, systemic. Farmers learned through repetition, intergenerational transmission, and embedding knowledge into myth, ritual, and law. The earliest "laboratories" were the fields; the earliest "archives" were the collective memory of agricultural communities. The genius lay not in isolated techniques but in their integration, forming a resilient ecosystem that sustained millions without synthetic inputs.

These achievements are not merely historical curiosities; they offer alternative logics of innovation. Modern industrial agriculture relies on chemical inputs—fertilisers, pesticides, herbicides—producing high yields at enormous ecological cost: soil depletion, water contamination, biodiversity loss. Ancient methods suggest a different approach: innovation through observation, resilience through alignment with natural cycles rather than domination. Incan terraces, for instance, distributed risk across microclimates; a crop failure at one level might be offset by success at another. Modern monocultures, by contrast, amplify vulnerability.

Crucially, these agricultural strategies demanded not only technical skill but cultural imagination. To domesticate a plant is to re-envision its potential; to cut a canal is to redefine a river; to mark the heavens is to render time calculable; to build a granary is to assert that the future can be secured. Innovation was conceptual before it was material. Ancient farmers were theorists of ecology, astronomers of the practical sky, engineers of time. Their work laid the intellectual foundations not merely for food systems but for civilisation itself.

This chapter unfolds these dimensions—domestication, irrigation, calendars, storage—not as separate tales but as facets of a single challenge: feeding large populations with limited tools and capricious environments. Each innovation demonstrates the power of human intelligence, patience, and organisation. Domestication shows how humans reshaped nature through careful experiment; irrigation reveals collective governance on vast scales; calendars expose the union of cosmology and ecology; storage illustrates mastery over uncertainty. Together, they exemplify ingenuity achieved without chemistry, a testament to what disciplined minds and coordinated communities can accomplish.

These systems were not perfect. Salinisation threatened Mesopotamian fields, dietary reliance on grain could narrow nutrition, and granaries could not always prevent famine. Yet even their flaws underline a vital truth: feeding millions has never relied solely on tools or inputs. Success demanded knowledge, resilient strategies, and cultural frameworks that imbued labour with purpose. The remnants of canals, terraces, temples, and granaries continue to speak, testifying to sophisticated systems of survival forged through intellect as much as sweat.

To nourish millions without industrial chemistry was not a miracle of abundance but a triumph of thought. Ancient farmers cultivated cooperation with plants, animals, rivers, skies, and time itself. This chapter examines how those inventions functioned, how they sustained societies, and what they reveal about human ingenuity. In the story of early agriculture lies not only the foundation of civilisation but lessons for today: insights into resilience, sustainability, and the enduring power of knowledge applied patiently across generations.

Domestication as Invention

The act of turning wild species into reliable partners was not an accident of nature but a deliberate, generational experiment. Few things appear more ordinary to us than the wheat in bread or the cow in a pasture, yet each is the product of human design no less than a bridge or a machine. The grasses that became wheat once shattered their seeds easily, scattering them beyond collection; early varieties of maize were scarcely more than a handful of kernels on a stalk; wild cattle were dangerous and unpredictable. To transform them into dependable sources of food and labour required centuries of observation, patient selection, and a memory system that could retain lessons across lifetimes. Domestication was invention in its most radical form: not the creation of new matter but the reshaping of life itself, guided by human thought long before the advent of formal science.

Selective breeding was not labelled as such, but it was unmistakably practiced. Archaeological evidence shows that by 8000 BCE, communities in the Fertile Crescent were already favouring cereal strains whose seeds remained attached to the stalk rather than falling to the ground. The act of gathering, winnowing, and resowing these particular seeds imposed a human filter on evolution, a conscious preference that altered the genetic profile of the plant. The same principle held true in animal management. From the aurochs, a massive and aggressive wild ancestor, humans gradually coaxed cattle that could be tethered, milked, and eventually yoked to ploughs. Such transformations required not only physical handling but the ability to anticipate traits and to value potential that might only emerge after many reproductive cycles. *This was not happenstance but strategy, a long experiment in shaping the biological world to suit cultural needs.*

What distinguishes domestication from mere use of wild species is intentionality. Hunting and gathering could sustain communities, but it did not involve altering the animals or plants themselves. Domestication, by contrast, is a creative act—identifying traits worth perpetuating and discarding others. It is telling that early farmers often preserved not just food species but symbolic or companion animals, from dogs trained to hunt cooperatively to cats that adapted to human settlements. This duality suggests that the impulse to domesticate was both practical and cognitive, tied to survival but also to an understanding of relationship and control. It reflects a recognition that humans could intervene in the processes of heredity, long before the language of genetics existed.

The experimental nature of domestication is best illustrated by maize. Its ancestor, teosinte, is a spindly grass native to Mesoamerica, bearing tiny cobs with a few hard kernels encased in tough shells. Yet by the first millennium BCE, Mesoamerican farmers had produced maize varieties with large, soft, tightly packed kernels that could sustain whole populations. The transformation is so extreme that maize cannot survive in the wild—it relies entirely on human cultivation for reproduction. This degree of dependence underscores the mutuality of invention: just as humans shaped maize, maize shaped human society, anchoring settled villages, ceremonial centres, and empires. *It is a reminder that invention can be reciprocal: in domesticating species, humans themselves became domesticated to rhythms of sowing, harvesting, and stewardship.*

Domestication also relied on patience that can be difficult to fathom in an age of immediate results. Unlike mechanical inventions, where a tool might be crafted and tested within days, breeding experiments unfolded across decades. The farmer who chose seeds from the most promising stalk would not see the full impact in their lifetime. They were participating in a transgenerational project, one that demanded faith in continuity and confidence that their descendants would recognise and build upon their work. This form of invention highlights an alternative logic of science—incremental, cumulative, and embedded in community memory rather than isolated brilliance. *It shows that some of humanity's most enduring technologies were created not by flashes of insight but by centuries of iteration.*

The very idea of selective breeding was not confined to plants and animals but was extended conceptually to human society itself. Ancient texts reveal metaphors of lineage, purity, and refinement that echo agricultural practice. When Sumerian scribes described dynasties as "seed" and "stock," they were drawing directly from the agricultural vocabulary of controlled reproduction. This demonstrates how domestication influenced not only diets but thought-worlds, embedding notions of improvement, control, and transmission into social structures. In this sense, agriculture was both a technical and a philosophical revolution.

The risks of domestication were equally profound. Early reliance on a few staple crops created vulnerabilities to drought, disease, or pest infestations. The Irish famine of the nineteenth century, though much later, illustrates how dependence on a single genetic line can lead to catastrophe. Ancient farmers were not oblivious to this danger. Evidence from Andean agriculture shows deliberate planting of diverse potato varieties across different altitudes, a strategy to hedge against environmental unpredictability. Such practices reveal that domestication was not blind optimisation for yield alone but also a conscious balancing of risk and resilience.

Domestication was also tied to ritual and identity. The selection of seeds was often accompanied by ceremonies, prayers, or offerings, embedding technical choices within cosmological frameworks. Far from being distractions, these rituals ensured continuity, transmitting knowledge through memory and meaning. They reinforced the value of careful selection and instilled discipline in agricultural communities. *Rituals were not merely symbolic; they were cultural mechanisms that safeguarded the integrity of experimental practices across generations.*

The intellectual depth of domestication becomes clearer when contrasted with modern scientific breeding. Today, laboratories can sequence genomes, isolate genes, and engineer traits within years. Yet the logic remains recognisable: choose desirable traits, propagate them, and stabilise the lineage. What differs is not the principle but the speed and tools. This continuity should caution us against dismissing ancient farmers as "primitive." Their methods, though lacking in chemical knowledge, achieved outcomes that remain foundational to human civilisation. Every loaf of bread and every glass of milk traces its lineage back to their patient experiments.

What makes domestication so compelling as an invention is that it demonstrates innovation without machines. It shows that thought, discipline, and intergenerational knowledge are themselves technologies. The modern concept of "biotechnology" may conjure images of laboratories and sterile equipment, but the first biotechnology was a farmer's hand choosing seeds from a sheaf. The results were as transformative as any industrial revolution: they enabled permanent settlements, population growth, and the rise of complex societies. Without domestication, there would be no cities, no writing, no history as we know it.

Seen in this light, domestication stands as one of humanity's greatest intellectual achievements, achieved without written theories or formal institutions of science. It exemplifies how experimentation can be embedded in daily life, how knowledge can be accumulated across generations, and how invention can take the form of patience rather than speed. Above all, it illustrates that the shaping of civilisation begins not only with monuments or tools but with seeds and herds, with the quiet but relentless pursuit of reliability in an unreliable world. The science of domestication was not written in books but in fields and flocks, a testament to the power of observation, memory, and imagination sustained across centuries.

Water Logic

Civilisations have always risen or fallen according to their relationship with water. Unlike stone or metal, which remain fixed once shaped, water resists control: it floods, evaporates, seeps into soil, and erodes boundaries. To direct it requires more than force; it requires foresight, cooperation, and design. What makes ancient irrigation remarkable is not simply its engineering but the *thinking* behind it—a logic that sought not only to move water but to reconcile human needs with the rhythms of rivers, rainfall, and terrain. It was, in many respects, one of humanity's earliest large-scale experiments in systems management, a synthesis of observation, mathematics, and politics sustained across centuries.

The Mesopotamian plain offers perhaps the clearest early example of this interplay between necessity and ingenuity. Situated between the Tigris and Euphrates, its fertile soils were paradoxically uninhabitable without intervention. Seasonal floods alternated with long periods of drought, leaving the land unpredictable. From around 6000 BCE, communities began to dig canals to channel river water into fields, using sluice gates and embankments to regulate flow. These were not crude trenches but carefully calculated systems requiring labour mobilisation and maintenance schedules. Clay tablets from Sumer record disputes over water rights, fines for neglecting ditches, and rules for proportional distribution—evidence that irrigation was as much a legal and administrative invention as a physical one. *Managing water meant managing society itself, embedding law into landscape.*

What makes Mesopotamian irrigation striking is its dependence on collective discipline. A canal benefits everyone it passes, but only if it is regularly dredged and repaired. If one farmer failed to maintain his section, entire downstream communities might lose access. This created a structural incentive for governance: kings, temples, and later bureaucracies justified their power in part by claiming authority over water. Irrigation, therefore, was not merely technical infrastructure but a foundation of political legitimacy. The earliest cities—Uruk, Lagash, Babylon—were not just concentrations of people but nodes in a hydraulic network. The capacity to store and redirect water determined not only yields but hierarchies, law, and urban form.

While Mesopotamia illustrates the birth of irrigation bureaucracy, the Nile Valley reveals a different logic: adaptation to predictability. Unlike the erratic floods of the Tigris and Euphrates, the Nile inundated with remarkable regularity, swelling each year from rains in the Ethiopian highlands. Egyptian farmers responded not by building massive canals but by devising basin irrigation. Fields were enclosed by low mud walls that trapped floodwater when it rose and released it slowly as it receded. This system aligned with the natural pulse of the river rather than attempting to override it. Egyptian kings claimed divine authority over the flood, but in practice, their genius lay in synchronising social calendars, tax systems, and agricultural cycles to a water pattern that was already regular. The annual inundation became not only an agricultural event but a cosmic one, tied to astronomy and religious ritual. *Where Mesopotamian irrigation demanded constant correction of unpredictability, Egyptian water logic exemplified calibration to rhythm.*

Across the world, in the Andes, an entirely different challenge presented itself. The highlands were not threatened by floods but by scarcity, steep terrain, and irregular rainfall. The Inca response was terraces carved into mountainsides, supported by stone walls and underlain with layers of gravel and soil designed to capture, filter, and retain water. Channels diverted glacial melt into these terraces, distributing it with astonishing precision. The system required no pumps, no metal pipes, only gravity and meticulous observation of slope and soil. Terraces extended arable land into otherwise barren slopes and reduced erosion while maintaining moisture even in dry seasons. Such designs demonstrate an intuitive grasp of hydrology and ecology: *water was not forced across landscapes but persuaded to linger.*

The ingenuity of the Inca lay not only in construction but in institutional memory. Local communities, or ayllus, were entrusted with maintaining particular terraces and channels, embedding water management into social organisation. Ritual offerings to mountain spirits (apus) reinforced the discipline of upkeep, aligning cosmology with maintenance schedules. In this way, technical systems were never divorced from cultural frameworks—they endured because they were *meaningful* as well as functional. This duality is a hallmark of sustainable infrastructure: it survives not by engineering alone but by being woven into the social fabric.

Contrasting traditions show how different environments demanded distinct forms of water logic, yet a common intellectual thread runs through them: the capacity to think in terms of flow, balance, and long-term consequence. This distinguishes irrigation from simple water collection. A well satisfies an individual household, but a canal or terrace system requires planning for generations. Decisions about sluices, reservoirs, and crop rotations had delayed consequences, demanding an understanding of cause and effect across time. In this sense, irrigation systems embody one of the earliest recognisable forms of scientific reasoning: hypothesis (if water is directed here, yields will rise), experiment (construct the channel), observation (measure the harvest), and adjustment (widen or narrow the gate).

What differentiates this from later laboratory science is its scale—it was not confined to controlled experiments but enacted across landscapes.

The sustainability of ancient systems often depended on how well communities accounted for unintended consequences. In Mesopotamia, over-irrigation gradually led to salinisation of soils, a problem still visible in the archaeological record. Crops declined, forcing shifts to more salt-tolerant varieties such as barley. Some scholars argue that this ecological feedback loop contributed to political instability, as declining yields undermined state authority. The lesson is clear: *water management carries risks of overconfidence; control can turn into collapse if cycles of replenishment are ignored.* In contrast, Andean terrace systems demonstrate remarkable resilience, with many still in use centuries after Spanish conquest, suggesting a design philosophy more attuned to ecological limits.

Modern parallels make these insights more than historical curiosities. Today, cities face challenges of water scarcity, flooding, and overuse of aquifers. Contemporary engineers deploy satellites, sensors, and desalination plants, yet the principles underlying sustainability remain ancient: distribute fairly, avoid over-extraction, design with the landscape, and align governance with natural cycles. Singapore's rainfall capture, Israel's drip irrigation, or California's contested water rights echo the same dilemmas faced by Sumerian farmers or Inca engineers. The difference lies in scale and technology, not in the underlying logic. *Water demands systems thinking, not just mechanical solutions.*

What gives ancient irrigation its enduring significance is that it was both technological and philosophical. To build a canal or a terrace was to ask: How do we live with water's unpredictability? How do we coordinate effort across communities? How do we align human cycles with natural ones? These are not only agricultural questions but existential ones, shaping the very structures of governance, belief, and identity. The fact that such systems were sustained for millennia testifies to the depth of their reasoning. Water was never simply a resource; it was a teacher, obliging societies to think collectively, long-term, and in dialogue with the land.

By looking at irrigation as an intellectual achievement rather than a mere feat of engineering, we see that ancient societies were not simply "making do" with rudimentary tools but developing systems of extraordinary sophistication. Their logic—patient, systemic, and ecological—offers a counterpoint to the short-term efficiencies of modern extraction. It suggests that sustainability is less about advanced chemistry or machinery than about cultivating the right habits of thought. If stone and bronze technologies reshaped matter, irrigation reshaped minds, teaching humans to think in terms of flow, interdependence, and the delicate art of balance.

Calendars as Agricultural Tools

The first calendars were not designed for philosophers, priests, or kings but for farmers. Long before written history, communities sought patterns in the sky to bring order to the uncertainties of the earth. A single mistimed sowing could mean famine, while the correct alignment of seed with season ensured survival. For this reason, astronomy and agriculture developed together: every marked solstice, every tracked star, was less an abstract pursuit than a calculation of when to plough, plant, and harvest. The calendar was humanity's earliest bridge between cosmology and ecology—a device that translated celestial regularities into agricultural security.

Archaeological evidence suggests that awareness of seasonal cycles predates urban civilisation by millennia. The monument of Nabta Playa in southern Egypt, dating to around 5000 BCE, contains stone alignments marking the summer solstice, a critical cue for anticipating the annual Nile flood. Similarly, the megaliths at Stonehenge are not random circles of stone but a solar and lunar observatory calibrated to track solstices and eclipses. These structures reveal something profound: *observation of the heavens was an agricultural instrument as much as a spiritual act.* By inscribing the sky into stone, early communities ensured that knowledge of planting cycles could be transmitted across generations, resisting the vagaries of memory or oral tradition.

Mesopotamia refined this impulse into written calendars. By 2000 BCE, Babylonian astronomers recorded celestial events with systematic precision. They noted that the rising of certain constellations coincided with seasonal changes—most famously, the heliacal rising of Sirius, which in Egypt reliably heralded the Nile flood. These observations allowed priests and scribes to create almanacs that structured farming schedules. Planting barley, for example, was not left to intuition but to calculations derived from the heavens. The calendar was thus both a scientific and political tool: whoever controlled it controlled food, and by extension, society itself. Kings could levy taxes and mandate labour because they could predict when harvests would come. *Time became a resource to be managed with as much care as water or soil.*

The Mayan civilisation illustrates the sophistication that calendrical thinking could achieve. Their 260-day ritual calendar, the *Tzolk'in,* interlocked with the 365-day solar calendar, the *Haab',* creating a 52-year "Calendar Round" that integrated cosmology with agriculture. The Mayans tracked the positions of Venus with extraordinary accuracy, not out of idle curiosity but because they linked the planet's cycles to rainfall patterns and maize planting. Fields were sown not simply when conditions looked favourable but when the calendar ordained it. Such precision was not superstition but empirical pattern recognition, grounded in centuries of observation. Even if the cosmological framework seems alien to modern sensibilities, the underlying practice reveals a disciplined form of generational science: *systematic correlation between celestial cycles and ecological outcomes.*

The Inca, though lacking a written script, developed a similarly nuanced temporal system. In Cusco, pillars known as *sucancas* were erected on horizons to mark the sun's rising and setting points across the year. These stone markers enabled priests to predict solstices and equinoxes with striking accuracy, coordinating planting and harvesting across a vast empire of diverse climates. The fact that these markers were public monuments, visible to all, demonstrates how deeply agricultural time was embedded into social space. The calendar was not hidden in scrolls or the hands of elites alone; it was carved into the landscape, ensuring collective alignment with natural cycles.

Yet calendars were more than scheduling devices; they embedded ecological awareness into cultural identity. Agricultural societies needed to anticipate not just when to sow but when to store, when to prepare for lean seasons, and when to mobilise labour. By fixing these cycles to celestial events, communities ensured that agricultural tasks were synchronised across households and regions. This synchronisation was essential for irrigation maintenance, pest control, and collective harvesting. Without calendars, such coordination would have been haphazard. With them, societies could scale agriculture to feed cities and sustain empires.

There is a philosophical insight here worth underlining: *to measure time is to domesticate uncertainty.* Nature offers no calendar; it offers rhythms, but they are irregularly felt. Some years floods come late, rains fail, or frosts arrive early. By tethering agriculture to astronomy, humans found a reference point beyond the local and the unpredictable. The sun would rise at the solstice regardless of drought or storm; the moon would wax and wane even if rivers ran dry. This gave farmers a stable framework against which to plan in the face of ecological variability. Calendars, then, were not simply records of what would happen but anchors of trust in the possibility of order amidst flux.

Of course, the very success of agricultural calendars depended on their adaptability. The Roman calendar, for instance, famously drifted out of alignment with the solar year, leading to sowings and festivals occurring at the wrong time of year. Julius Caesar's reform in 46 BCE, creating the Julian calendar, was not merely a bureaucratic tidying of dates but a corrective to agricultural disorder. Without recalibration, harvests risked misalignment with seasons. Later refinements under Pope Gregory XIII in 1582, producing the Gregorian calendar still used today, were again motivated by concern with solar accuracy, since even minor drifts could undermine long-term agricultural reliability. These reforms remind us that timekeeping is not static but a constant negotiation between celestial observation and human need.

Modern agriculture still bears the imprint of ancient calendrical thinking, even if cloaked in new forms. Farmers in many regions continue to rely on lunar cycles, planting certain crops on waxing or waning moons, a practice supported by empirical evidence that gravitational pull affects soil moisture. Meanwhile, climate science has introduced long-range forecasting models that resemble an updated form of agricultural astronomy. Satellites now track El Niño patterns, offering predictions that inform planting strategies across continents. Yet the principle is unchanged: *align human action with predictable cycles of nature to reduce risk and maximise yield.* What differs is that our calendars are now digital and global rather than carved in stone or woven into ritual.

At the same time, our distance from the sky has consequences. Urban populations, lit by electric lamps, often lose direct contact with celestial cues that once structured daily life. Few today know when Sirius rises or can distinguish a waxing from a waning moon. This detachment does not eliminate dependence—it obscures it. The food that reaches supermarket shelves still depends on rhythms of sun, rain, and season, but the calendrical discipline behind it is hidden in supply chains and data models rather than visible monuments or communal rituals. The risk is that this abstraction dulls awareness of ecological limits, making agriculture seem like an industrial process rather than a dialogue with planetary cycles.

Ancient calendars remind us that agriculture is not just about soil and seed but about *time itself as a medium of cultivation.* To farm well requires not only controlling land and water but understanding when the earth is ready to receive and when it must be allowed to rest. The sky provided the reference frame for this understanding, embedding agricultural practice within a broader cosmology that linked human labour to cosmic order. Whether through Sumerian tablets, Egyptian star charts, Mayan glyphs, or Incan stone pillars, the lesson is consistent: sustainable farming begins with the ability to measure, anticipate, and coordinate time. The calendar was not an auxiliary tool to agriculture; it was agriculture's intellectual backbone, ensuring that cosmology and ecology were never conceived as separate domains.

In recognising this, we can appreciate that the oldest sciences were never purely about the heavens or purely about the earth, but about the *relationship between the two.* Astronomy and agriculture were twin disciplines, each meaningless without the other. The calendar, in this sense, was less a record of days than a contract between humanity and nature, binding survival to the discipline of observation. That contract, forged thousands of years ago, still underwrites every loaf of bread and bowl of rice today.

The Mathematics of Storage

The most revolutionary invention in the history of agriculture was not the plough, nor even irrigation, but the granary. To harvest is only half of the problem; the true challenge is carrying food across the gap between seasons. Without preservation, abundance rots into scarcity, and harvest wealth becomes famine within months. The logic of storage is therefore a logic of time, transforming fragile crops into reliable sustenance. To build a granary is to engage in a kind of mathematics, measuring not only volume and weight but risk, duration, and decay. These were not abstract calculations but existential ones: survival depended on whether a community could transform perishable grain into enduring security.

The earliest granaries date back over eleven millennia to the Levant, where the Natufian culture constructed structures with plastered floors and raised platforms to keep rodents and moisture at bay. These were modest in size but immense in implication: the stored grain allowed people to remain in one place year-round, stabilising settlements and paving the way for agriculture to replace foraging. In effect, storage was the hinge on which civilisation swung. By extending food availability beyond the immediacy of harvest, these early granaries created a temporal surplus: time could be spent not solely on subsistence but on building, planning, and imagining. *The granary was humanity's first bank, its vaults filled not with coin but with calories.*

Mathematics quickly entered this equation. Once food could be stored, societies needed to measure precisely how much was kept, how long it would last, and how it should be rationed. The clay tablets of Mesopotamia, some of the earliest known forms of writing, are filled with tallies of grain deposits and withdrawals. Symbols evolved not from poetry but from the hard necessity of bookkeeping. The cuneiform wedge impressed into clay was a numerical tool as much as a linguistic one. An account of grain owed, stored, or distributed was not simply a record—it was a prediction of the future, an attempt to map supply across time. These early accountants were also agricultural mathematicians, calculating rates of consumption against the certainty of future scarcity.

Architecture itself became an expression of quantitative thinking. Egyptian granaries, often attached to temples, were designed with sloping floors to funnel grain efficiently and reduce spoilage. In the Indus Valley, massive brick structures in Mohenjo-Daro and Harappa, interpreted as granaries, reveal sophisticated ventilation systems. These were not ad hoc piles of surplus but carefully engineered storage devices, their design anticipating issues of humidity, pests, and spoilage. The mathematical challenge lay not just in counting grain but in controlling the variables of temperature, airflow, and volume that determined how long it remained edible. *To store food successfully was to turn physics into nutrition.*

The Inca, operating without wheels, written numbers, or beasts of burden, nonetheless mastered storage on an imperial scale. Their network of *qollqas*, stone storage facilities built at varying altitudes, exploited microclimates to preserve maize, potatoes, and quinoa. Grain stored at higher elevations lasted longer due to cooler temperatures and lower humidity. This was not guesswork but empirical optimisation: centuries of trial and observation encoded into infrastructure. Runners could move food across the empire with astonishing efficiency, ensuring that surplus from one region could buffer shortages in another. Here storage transcended the individual household, becoming a political technology that bound the empire together. Control over time—by controlling grain—translated directly into control over people.

Philosophically, storage represents a profound shift in the human relationship with uncertainty. For hunter-gatherers, risk was externalised: survival hinged on the presence of game or the ripening of wild fruits. Agriculture, by contrast, internalised risk management. Through storage, societies could buffer themselves against the caprices of weather or pests. Yet this was never foolproof. Rats, mould, and theft were as threatening as drought. The mathematics of storage was therefore also a mathematics of loss. How much spoilage was acceptable? How much surplus should be reserved for unforeseen disasters? These questions forced communities to quantify risk long before the formal language of probability emerged. The granary was both a safeguard and a wager: a bet that the volume stored would outlast the hazards waiting to consume it.

The cultural consequences of storage were equally transformative. Surplus enabled redistribution, taxation, and standing armies. Priests and kings rose not only because of divine charisma but because they could command stored grain. A warehouse of barley could be translated into power, loyalty, or military strength. In this sense, storage was the first political economy, a system in which the mathematics of granaries structured hierarchies and obligations. When Joseph in the Hebrew Bible advised Pharaoh to store grain during years of plenty, it was a continuation of this logic: preservation as a political act, a demonstration that command over food equalled command over destiny.

Preservation techniques amplified these effects. Smoking, drying, fermenting, and salting were forms of applied science long before chemistry named them. Each technique was a manipulation of time, slowing decay through environmental control. Fermentation in particular reveals a sophisticated understanding of transformation: allowing microbes to alter food in ways that made it both safer and longer lasting. Beer and bread, staples of early civilisations, were as much technologies of preservation as of nutrition. These practices required not only empirical skill but cultural discipline: rituals and taboos ensured that preservation methods were followed faithfully, embedding practical science within symbolic frameworks.

Modern parallels show how deeply these ancient calculations persist. Refrigeration and vacuum packaging are today's equivalents of the raised-floor granary or ventilated storehouse. Supply chain logistics, governed by algorithms, still wrestle with the same central question: how to move food through time without letting it spoil. The mathematics is more advanced, incorporating differential equations and climate models, yet the underlying challenge is recognisable to a Sumerian scribe tallying grain in a clay storehouse. In fact, the global problem of food waste—where up to a third of all food is lost between farm and table—echoes ancient anxieties about spoilage. What has changed is not the question but the scale: *our granary is now planetary.*

Critically, storage also shapes psychology. Knowing that food will be available tomorrow alters how a society thinks about labour, planning, and even morality. Surplus makes saving and lending conceivable; it allows art and philosophy to flourish because not everyone must farm. Yet it also introduces inequality: those who control storage control others' futures. In this way, the mathematics of storage is not merely arithmetic but ethical. How much should be shared? Who deserves access? These questions still animate debates about food security today. Modern stockpiling of grain reserves, or controversies over the strategic hoarding of rice and wheat by powerful nations, are direct descendants of dilemmas first confronted in mud-brick silos and stone granaries.

What is striking is how storage reveals the continuity between ancient pragmatism and modern science. Both recognise that time is the greatest adversary of food, and both respond with quantitative ingenuity. Ancient engineers aligned airflow with seasonal humidity; modern logistics experts align shipping schedules with global consumption patterns. Both engage in a mathematics that is less about numbers on a page than about shaping probabilities in the service of survival. The lesson across centuries is unmistakable: *to master storage is to master time itself.* And in doing so, humanity secured not only its meals but its future.

Chapter 5
Weapons of the Mind

The story of human conflict is often told in the clang of bronze, the thunder of hooves, or the flash of iron blades. Yet beneath these visible instruments lies a quieter, more enduring force: the intellectual innovations that made violence systematic, precise, and, at times, devastatingly effective. Battlefields have long served as testing grounds not merely for courage but for thought, where geometry, psychology, architecture, and philosophy converge to decide survival or annihilation. To speak of the weapons of the mind is to acknowledge that many decisive victories hinged less on muscle than on the capacity to reason—mathematically, strategically, and ethically—about destruction itself.

Take the humble arrow. Firing it is a simple act, but the crafting of a bow—calculating limb curvature, string tension, and flex—required an intuitive grasp of geometry. Ancient engineers did not solve equations on parchment, yet their experimental refinements rival the precision of modern design laboratories. The Greeks transformed parabolic flight into timber, rope, and iron with their catapults and ballistae. The Chinese, centuries earlier, perfected crossbows with sophisticated locking mechanisms, harnessing leverage and stored energy to achieve remarkable accuracy. In every instance, it was the mind, rather than the hand, that became the true weapon, converting raw strength into measured projection. Mathematics, once embodied in timber or sinew, became an engine of war.

Yet warfare's intellectual arsenal extended beyond numbers. Equally powerful was the manipulation of the unseen: fear, confusion, and awe. Armies have always been accompanied by spectacle. The Assyrians, notorious for their cruelty, employed terror deliberately, decorating palaces with the flayed and the defeated to intimidate future foes. Greek hoplites advanced not silently but in a rhythmic thunder of shields and chants, turning psychological pressure into tactical leverage before the first spear was thrown. Later, Mongol cavalry perfected feigned retreats, weaponising unpredictability to devastating effect. Commanders understood that the mind, unsettled, yields more readily than the body. The manipulation of perception—what an enemy sees, hears, and expects—often proved more decisive than the sharpest blade.

Defence, too, demanded intellectual rigour. Fortresses were not random accumulations of stone; each wall, tower, and moat represented a hypothesis about resisting violence. Jericho's ancient walls reveal the early leap from individual to collective protection. By the classical period, cities such as Syracuse and Constantinople were encircled by layered defences, designed not merely to keep invaders at bay but to shape their behaviour, slowing, exposing, and exhausting them. Medieval castles extended this logic, exploiting space itself with concentric rings, kill zones, and complex geometry. Behind every wall stood a mind considering space as a field to manipulate, anticipating moves and counter-moves long before modern game theory had a name.

Yet every advance carried a moral shadow. Innovation in weapons or fortifications raised questions of justification. How much ingenuity should be devoted to destruction?

Designing a catapult or raising a fortress presupposes a philosophy of violence, one that had to be reconciled with the society that built and endured it. Some cultures celebrated it, exalting gods of war and divine sanction for conquest. Others tempered their creativity with reflection: the Mahabharata teems with agonised debate over the ethics of using superior weapons against the weak. Romans framed military engineering as the enforcement of order, yet writers frequently acknowledged unease at the devastation their legions inflicted. These philosophical tensions endure, reminding us that to innovate in war is always also to confront moral responsibility.

The intellectual legacy of ancient warfare persists far beyond decayed battlements and rusted catapults. Modern ballistics, with algorithms and simulations, descends directly from the geometric reasoning that guided the first bows. Psychological warfare, now propagated through media and cyber campaigns, traces its principles to drummers, banners, and battlefield spectacle. Even ethical dilemmas echo earlier debates: nuclear deterrence mirrors the uneasy realisation that human ingenuity, once harnessed to destruction, can outstrip control. Across centuries, the same principle holds: war is a crucible of thought, a harsh laboratory in which ideas are tested under pressure.

This chapter is not a catalogue of weapons but an exploration of how intelligence itself became militarised. The bow and the fortress, the war drum and the siege tower, are not merely artefacts of conflict; they are monuments to human cognition under extreme conditions. They reveal how communities reasoned about trajectory, perception, space, and justice when survival hung in the balance. More provocatively, they demonstrate that the mind can itself be weaponised, honed not for creation or understanding but for the precise calculus of domination. Every fortress raised invited a siege; every engineered weapon, escalation.

What follows is an inquiry into these weapons of the mind: the geometry that guided arrows and catapults, the psychology that converted sound and spectacle into terror, the architectural reasoning that transformed stone into bastions, and the moral reflection—or its absence—that framed the use of engineered violence. Understanding these is not an endorsement, but a recognition that innovation and responsibility are inseparable, and that the human capacity to think has always been as lethal as it is creative.

Geometry of the Bow and Catapult

The arc of a weapon is also the arc of an idea. The first time a hunter bent a branch into tension and released a stone-tipped arrow, he was not only unleashing force but discovering geometry in practice. What appears simple—a string pulled back, a shaft released—was in truth an experiment in curvature, elasticity, and trajectory. Every bow ever built, from the simplest self-bow to the intricately laminated composite bows of the steppes, embodied implicit calculations about angles, leverage, and stored energy. Before mathematics was written in symbols, it was inscribed in wood and sinew.

The bow is perhaps the earliest tool in which mathematical reasoning became inseparable from survival. Its efficiency relies on the careful balance of force and distance: the longer the draw, the greater the potential energy stored, but also the greater the strain on the bow's materials. Early bowyers may not have used formal equations, but through centuries of iteration they mastered a kind of experimental geometry. They learned that a straight stave would yield limited power, while a recurved or reflexed design amplified the energy stored within the limbs, producing arrows that flew farther and struck harder. This was a lesson in curves—an intuitive grasp of the fact that altering the geometry of a weapon alters the mathematics of its performance. The composite bow perfected by nomadic horse archers, layering wood, horn, and sinew, was not only a feat of craftsmanship but a mathematical triumph: compact enough to wield on horseback, yet powerful enough to penetrate armour. It redefined the balance between size and force, an equation that shaped empires.

Equally revealing is the transition from handheld weapons to large-scale engines of war. When Greek and Roman engineers began developing torsion-powered catapults, they were in effect formalising the geometry of the bow into scalable systems. The catapult is a bow multiplied: its sinew ropes twisted into tension, its wooden arms extending the leverage of muscle into magnitudes of force no individual could summon. Designing such machines required attention to ratios and angles. How thick should the sinew bundle be to maximise stored energy without snapping? At what arm length could force be sustained without diminishing speed? These were not abstract questions. A miscalculation in angle or proportion could cause catastrophic failure, wasting resources or costing lives. The mathematics of the catapult was unforgiving, and its refinement became one of the earliest intersections of science and warfare.

The Greeks approached this challenge with a systematic rigour. By the fourth century BCE, under the influence of mathematicians like Archytas and engineers such as Ctesibius, siege engines were being designed with calculated precision. The surviving *Belopoeica* of Philon of Byzantium outlines formulae for scaling catapults according to the weight of projectiles. Here, mathematics became codified as military doctrine, transforming trial-and-error into replicable engineering. A stone-thrower was no longer built by instinct alone but calibrated according to numerical relationships between cord diameter, arm length, and projectile mass. *Geometry had become not merely descriptive but prescriptive: it dictated how power could be constructed and projected at will.*

The Romans inherited and expanded this knowledge, embedding catapults and ballistae into their legions as standard equipment. The ballista in particular reveals the union of mathematical thinking and military pragmatism. Essentially a giant crossbow, it required precise carpentry to ensure the twin arms rotated symmetrically under torsion, launching bolts with remarkable accuracy. Unlike the brute force of the onager, the ballista was valued for its precision, capable of targeting enemy officers or striking fortifications at calculated ranges. Roman manuals specify angles of elevation, adjustments for wind, and corrections for distance, demonstrating a culture that understood war not only as violence but as a geometry problem to be solved.

One of the most fascinating aspects of these weapons lies in their role as laboratories of mathematical thought. The parabolic flight of a projectile is, in modern terms, a problem in physics. To the ancients, it was a problem in practice: how to adjust the release angle to extend range, how to alter projectile mass to increase penetration, how to account for the resistance of air or the slope of a battlefield. Generals and engineers alike became students of trajectory. In Syracuse, Archimedes is said to have designed war machines that hurled stones with astonishing accuracy, employing his understanding of levers and pulleys to magnify human strength. Whether or not all accounts are historically accurate, the legend itself testifies to a broader truth: *the battlefield was a classroom, and geometry its curriculum.*

Yet the application of mathematics to weapons was never straightforward. There was always a tension between the purity of abstract reasoning and the messiness of material reality. A formula might specify the ideal cord tension, but local humidity could weaken sinew fibres. A calculated release angle might optimise trajectory in theory, but shifting winds rendered it imprecise in practice. Ancient engineers were thus forced to develop not only mathematical models but a pragmatic sense of tolerance and adjustment, blending calculation with craft. In this sense, their innovations prefigure modern engineering, where theory is always tempered by testing and iteration.

The spread of these technologies also illuminates the exchange of ideas across cultures. Chinese engineers, long before the Hellenistic world, had experimented with crossbows whose triggers incorporated complex mechanics of leverage and lock. Their development of repeating crossbows—firing multiple bolts in rapid succession—showed a different application of geometric reasoning, prioritising efficiency of release rather than maximum range. In India, the use of longbows alongside siege engines produced distinct tactical geometries on the battlefield, where the curve of a bow and the arc of a stone shared the same underlying mathematics of trajectory. These variations remind us that geometry was not universal in form but universal in principle: wherever humans sought to project force beyond their reach, they confronted the same questions of angle, tension, and arc.

The implications extend beyond ancient battlefields. The intellectual move from instinctive bow-making to formula-driven catapult construction is part of a larger story: the evolution of mathematics from lived experience to abstract science. What began as the bending of wood became a branch of mechanics; what started as empirical adjustments of angle became the study of parabolas. In this sense, war inadvertently accelerated intellectual progress. The need to hit a target with precision demanded reasoning that would later underpin astronomy, physics, and engineering. One might say that every stone hurled from a catapult was also a question hurled into the future, demanding an answer that mathematics would eventually provide in fuller form.

There is also a philosophical dimension to this history. The bow and the catapult demonstrate how human thought transforms nature's forces into extensions of intention. To bend wood is to capture elasticity; to twist sinew is to trap potential energy; to calculate a trajectory is to harness gravity itself as an ally. *The act of projecting force is also the act of projecting knowledge: an understanding of how the world works, channelled into power.* This duality—that knowledge liberates but also weaponises—remains central to the human story. The same geometric reasoning that guides a catapult also guides the trajectory of a spacecraft. The difference lies in purpose.

Modern ballistics, with its equations and simulations, is a direct heir to these ancient experiments. The science of trajectory, now refined into algorithms, continues to apply the same principles: angle, force, resistance, distance. Yet the essential insight remains unchanged from the first bowyers and engineers: geometry governs projection, whether of an arrow or a missile. To study the bow and catapult is therefore to study the origins of applied mathematics itself—*a mathematics born not in quiet contemplation but in the urgent demands of survival and conflict.*

By examining these ancient weapons of projection, we uncover not only the ingenuity of early militaries but the deeper logic of human invention. In the curve of a bow and the arms of a catapult we glimpse the beginnings of systematic thought, the recognition that ideas have arcs as real as arrows, and that geometry, once harnessed, changes the balance of power. The bow taught humanity to extend the hand; the catapult taught humanity to extend the mind. Both stand as monuments to the capacity of mathematics to reshape the world—sometimes for survival, sometimes for conquest, but always as a testament to the restless intellect behind the string and the stone.

Psychology of Fear

A soldier on the battlefield is not merely a body carrying weapons; he is a mind interpreting sound, sight, and uncertainty. The collapse of an army often precedes the clash of steel, beginning in the imagination rather than the flesh. Ancient war leaders understood that terror, once seeded, spread faster than fire, and that the surest way to weaken an opponent was not always through direct killing but through the deliberate manipulation of fear. *To unsettle the mind was to win half the battle before the first strike was made.*

Sound was among the earliest and most potent tools of psychological warfare. Consider the horns of the Assyrians, whose armies advanced with blasts designed not only to coordinate movements but to overwhelm their enemies with a sense of encroaching inevitability. The sound of drums and trumpets amplified the perception of numbers, making a marching column seem like an unstoppable tide. In China, bronze war drums achieved similar effects, their thunder echoing across valleys to create the illusion of vast, coordinated forces. Such instruments were not ornamental; they were carefully chosen for their resonance, their capacity to travel distance, and their ability to shape perception. *Fear is never just the recognition of danger—it is the anticipation of what might follow, magnified by uncertainty.* Sound worked precisely because it blurred the line between reality and imagination.

Equally effective was spectacle. The ancient Celts, for instance, entered battle naked, their bodies painted in vivid colours. To the Romans, who prized order and discipline, this display was unsettling: it inverted the expectation of warriors armoured for protection, replacing it with a spectacle of fearless defiance. Likewise, war chariots—thundering across plains with wheels clad in iron—created sights as terrifying as the weapons they carried. Horses adorned with armour or fire-breathing devices, as in Byzantine Greek fire tactics, turned animals into apparitions of destruction. These performances were designed to short-circuit rational assessment in the enemy, substituting awe and dread for tactical calculation.

Unpredictability heightened this psychological effect. Fear thrives in uncertainty, and ancient commanders exploited this by developing tactics that defied expectation. Hannibal's crossing of the Alps is a famous example: the very improbability of his route disoriented the Roman psyche. They were not merely facing Carthaginian troops—they were facing an enemy who seemed willing to defy geography itself. Mongol armies later perfected the art of unpredictability, feigning retreat to lure opponents into ambushes. Their reputation became so formidable that cities often surrendered upon hearing of their approach, terror preceding their physical arrival. *Reputation itself became a weapon, forged through deliberate unpredictability and spectacle.*

There is a crucial distinction between the destructive power of weapons and the disruptive power of fear. A sword kills an individual; fear unravels a formation. Ancient tacticians recognised this difference and cultivated strategies to induce collective breakdowns. The flight of one soldier could ripple into the collapse of an entire unit, for fear is contagious. The Spartans, who trained their soldiers to resist panic through relentless drills, understood this contagion well. They forged cohesion precisely to guard against the psychological fractures that fear produced in less disciplined armies. Where one culture harnessed terror to destabilise, another cultivated mental resilience to resist it.

The ethics of fear are complex. On one hand, inducing psychological collapse often spared lives, as enemies fled rather than fought to the last. On the other, it represents the deliberate exploitation of human vulnerability. Was it more humane to terrify an enemy into retreat than to slaughter them through direct confrontation? Or was terror itself a cruelty that lingered beyond the battlefield, haunting survivors long after swords were sheathed? Ancient texts wrestle with these questions implicitly. The *Arthashastra*, the Indian treatise on statecraft, advocates the use of psychological tactics, including rumours and displays, as legitimate instruments of war. Yet Greek thinkers like Plato worried about the corrupting influence of deceit and spectacle, questioning whether warfare conducted through fear undermined civic virtue.

The persistence of such tactics into modern times underscores their effectiveness. Air raid sirens, the dropping of leaflets predicting destruction, or the use of sound weapons to disorient crowds are all heirs to ancient methods. The Nazi Blitz on London sought not merely to destroy infrastructure but to shatter morale through sound and unpredictability. The American use of "shock and awe" in the Gulf War was consciously rooted in the manipulation of perception: overwhelming displays of power designed to paralyse the enemy's will to resist. These modern echoes remind us that while weapons evolve, the human mind remains vulnerable to the same pressures as those faced on battlefields thousands of years ago.

Philosophically, the study of fear in warfare reveals a profound irony: the mind, humanity's greatest asset, is also its point of weakness. Rational thought, which enables innovation, can be overridden by primal responses to threat. The clash between cognition and instinct is central here. A soldier may know intellectually that the enemy's drums or chants do not increase their numbers, yet the visceral response—the pounding heart, the sense of smallness—undermines rationality. Commanders played upon this duality, using spectacle and sound to weaponise instinct against thought.

It is tempting to think of fear as a weakness, but history shows it was also a catalyst for innovation. The need to defend against psychological tactics spurred developments in training, ritual, and belief systems. Roman soldiers carried standards and emblems not only as identifiers but as anchors of morale. Religious rituals before battle, from sacrifices to omens, provided psychological armour, insulating the mind against uncertainty. Belief systems offered predictability where the battlefield did not, turning the chaos of war into a narrative where destiny rather than dread determined the outcome. *To fight fear, societies invented meaning.*

Understanding the psychology of fear in ancient warfare also illuminates broader truths about human behaviour. Fear is not irrational; it is hyper-rational, a mind attempting to predict outcomes when information is scarce. Its manipulation lies in the careful control of what is seen and heard, the deliberate engineering of uncertainty. In this sense, war innovators were early psychologists, probing the vulnerabilities of perception and exploiting them with artistry. They understood that the battlefield was as much an arena of minds as of bodies, and that a faltering spirit was often more decisive than a broken shield.

The legacy of these tactics invites reflection on the continuity of human psychology. Whether in ancient fields, medieval sieges, or modern cities, the instruments of fear—sound, spectacle, unpredictability—retain their force. What has changed is not the human mind but the scale and sophistication of its manipulation. By studying these origins, we recognise that warfare has always been a contest not merely of arms but of imaginations, a struggle to dominate not just land but the inner landscapes of hope and dread. Fear, in its ancient and modern forms, remains the most subtle and enduring weapon ever devised.

Fortresses as Thought Experiments

A fortress is not merely a wall of stone or earth; it is an argument in architecture, a material statement of how human beings imagine threats, time, and space. The defensive structures of antiquity reveal more than the anxieties of their builders—they show how entire societies conceived of risk, permanence, and the geometry of survival. To build a fortress was to engage in a kind of thought experiment, one where mathematics, psychology, and philosophy were as important as stone and mortar. The wall was not only about keeping an enemy out; it was about projecting foresight, broadcasting stability, and shaping the imagination of those within and those without.

The earliest city walls, such as those of Jericho dating back over 9,000 years, were colossal undertakings for communities that possessed only rudimentary tools. The scale of these constructions raises a fundamental question: why invest so heavily in barriers long before the emergence of large-scale professional armies? Archaeological evidence suggests that the walls of Jericho were not merely defensive but symbolic—a visible line between the known and the unknown, order and chaos. To encircle a community in stone was to define its identity. The wall was as much about shaping perception as it was about repelling assault. *A fortress always begins in the mind, as a line drawn against uncertainty.*

Yet once warfare grew more sophisticated, walls had to meet practical demands. The Assyrians, masters of siegecraft, deployed battering rams, siege towers, and scaling ladders. Their ingenuity forced defenders to adapt. High, smooth walls became inadequate once attackers learned to undermine foundations or build ramps. Cities such as Nineveh and Babylon answered with monumental ramparts, multiple layers of defences, and moats to stall siege engines. Each architectural feature was a calculation: How high must the wall be to resist ladders? How wide must the moat be to neutralise a ram? How many layers of defence could buy enough time for reinforcements to arrive? Defence became mathematics in stone, an engineering dialogue with the possibilities of attack.

The moat deserves particular attention as a paradoxical device. Unlike walls, which repel through obstruction, moats slow and frustrate. They exploit the fact that armies move with baggage, equipment, and supply trains, not as fluid streams of bodies. Crossing water, mud, or spiked ditches interrupts the rhythm of assault, buying defenders moments that might mean the difference between holding and collapse. In this sense, the moat was not a passive barrier but an active manipulation of time. It demonstrated an insight that would recur in many civilisations: victory in defence often depends less on absolute impenetrability than on the artful stretching of moments. *Fortifications are clocks as much as they are barriers.*

Nowhere was this intellectual arms race clearer than in the Mediterranean. Greek city-states developed acropolises, defensible high points that relied on elevation as a form of protection. The logic was as much psychological as physical: attackers confronted not only steep terrain but also the daunting image of a city literally above them, gazing down. In contrast, the Romans, inheriting and surpassing Greek methods, built walls as instruments of imperial order. Hadrian's Wall in Britain was less about fending off massive invasions than about asserting a line of control, a visible limit of Roman reach. Its genius lay in its dual function: a defensive barrier and an administrative tool, channelling trade, migration, and taxation. In this sense, fortresses extended beyond war—they became tools of governance, embodiments of how power organises space.

The medieval period refined the fortress as a system of layered defences. Castles in Europe, with their keeps, baileys, and curtain walls, were not random accumulations of stone but carefully orchestrated designs reflecting evolving ideas of siege warfare. As trebuchets and catapults became more powerful, walls thickened and adopted angled bastions to deflect projectiles. The very shape of fortresses shifted from simple enclosures to complex geometries. Every corner, tower, and gatehouse represented a hypothesis about how violence could arrive and how it might be resisted. These were experiments run not in laboratories but in landscapes, tested by armies and remembered in ruins.

The Renaissance transformed defensive architecture into an explicit science. The rise of gunpowder artillery shattered the medieval fortress, forcing engineers to rethink geometry itself. Italian military architects pioneered the trace italienne, or star fort, with its angled bastions and low, sloped walls designed to absorb cannon fire. These star-shaped citadels, like those still visible in Palmanova or Antwerp, represent perhaps the purest form of the fortress as a thought experiment. Their designers worked with precise calculations of ballistic trajectories, fields of fire, and overlapping kill zones. A fortress became a machine of prediction, a crystallised set of answers to questions about the range, angle, and speed of projectiles.

But fortresses also reveal cultural differences in how societies confronted risk. In China, the Great Wall embodied a philosophy of scale. Rather than fortify individual cities, dynasties sought to inscribe a barrier across thousands of miles, treating geography itself as part of the design. The logic was less about resisting a single siege and more about shaping the movement of entire peoples. Similarly, in Japan, castles such as Himeji incorporated not only walls and moats but elaborate psychological features: labyrinthine approaches designed to confuse attackers, create choke points, and exhaust morale. The defender's art was never merely technical—it was imaginative, anticipating not only physical assaults but also emotional states.

The thought-experimental nature of fortifications is most evident in their anticipatory quality. A wall imagines ladders. A moat imagines siege engines. A bastion imagines artillery trajectories. Fortresses are predictive structures, built not against present threats but against hypothetical futures. This forward-looking quality makes them philosophical artefacts as much as military ones. They embody a society's capacity to think beyond the immediate, to rehearse possible catastrophes and inscribe solutions into stone. In that sense, fortresses mirror other intellectual projects of their age, from legal codes to religious doctrines, which also sought to stabilise life against the unpredictabilities of human conflict.

The intellectual labour behind fortresses also carried a moral dimension. To wall oneself in is to make a statement about trust and exclusion. Walls divided civilisations from "barbarians," citizens from outsiders, wealth from deprivation. They reassured those inside even as they projected hostility to those outside. The ethics of fortification therefore extended beyond battlefields. Did such structures preserve life by deterring attack, or did they invite greater violence by escalating the logic of arms? Some societies saw walls as symbols of cowardice, arguing that true courage lay in meeting the enemy openly. Others embraced fortification as prudence, evidence of wisdom rather than weakness.

In the modern world, the fortress has not disappeared; it has evolved. Bunkers, fallout shelters, and even digital firewalls carry the same underlying logic: to pre-empt threat by constructing protective boundaries. The parallels are striking. Just as medieval engineers experimented with geometries to deflect catapults, today's cybersecurity experts experiment with algorithms to deflect intrusion. Both are engaged in the same fundamental act: rehearsing the enemy's mind and embedding answers into defensive design. The fortress, whether of stone or code, remains a thought experiment about vulnerability, ingenuity, and time.

Ultimately, the history of fortresses teaches us that defence is never static. Every wall is a hypothesis awaiting refutation by the next innovation in attack. What survives in the ruins of castles, star forts, and city walls is not only stone but the intellectual daring of those who imagined how the future might break upon them. *Fortifications are less about invincibility than about foresight; they are material philosophies of how to inhabit a dangerous world.*

Ethics of Destruction

The paradox of civilisation is that its greatest intellectual advances often coincide with its most sophisticated methods of killing. From the moment human societies began to master the engineering of violence—designing weapons, fortifications, and strategies that amplified destructive force—they also confronted a moral question: how far could they go without undermining the very values they sought to protect? The history of ancient warfare is not only a tale of ingenuity in metallurgy, mathematics, or tactics but also of attempts to contain, justify, or humanise the devastation that inevitably followed. *Every catapult, wall, and sharpened blade carried not only a technical design but a moral footprint.*

In Mesopotamia, some of the earliest recorded legal codes already grappled with the legitimacy of violence. The Code of Hammurabi did not merely regulate domestic disputes but also reflected the logic of war. To injure another unjustly invited proportionate punishment, establishing the principle that destruction must be balanced by justice. Warfare, however, was not so neatly circumscribed. Kings invoked divine sanction, claiming that gods endorsed their campaigns. Violence, in this framing, was not a matter of personal choice but of cosmic duty. The engineering of destruction was justified not as human cruelty but as divine necessity. *By placing bloodshed in the hands of gods, societies shielded themselves from the ethical burden of their own inventions.*

Yet even within such frameworks, restraint emerged. The Assyrians, infamous for their brutal sieges and psychological warfare, also developed elaborate rituals to legitimise conflict. War had to be preceded by omens and divinations. The enemy was demonised, their defeat framed as a purging of evil rather than a mere expansion of power. The very act of seeking celestial approval suggests that ancient rulers recognised a tension: unchecked violence, however effective, risked eroding the legitimacy of authority. The spectacle of destruction was permissible only if couched within a narrative that elevated it beyond mere conquest.

The Greeks brought a more explicit philosophical dimension to the ethics of destruction. While Sparta glorified martial discipline and collective sacrifice, Athens introduced debates about just war and proportionality. Herodotus and Thucydides chronicled campaigns where strategic necessity collided with moral ambiguity, most famously in the Melian Dialogue, where Athens argued that power dictated justice. This stark realism—that the strong impose their will because they can—sat uneasily alongside Greek ideals of reason and civic virtue. At the same time, philosophers such as Aristotle reflected on the role of warfare in achieving *eudaimonia,* or human flourishing, suggesting that violence could be legitimate only insofar as it secured conditions for a virtuous life. Here, the ethics of destruction became entwined with broader questions about the good life, shifting warfare from brute necessity to philosophical problem.

Rome, inheriting and systematising Greek thought, institutionalised restraint in the concept of *bellum iustum,* the just war. Before launching an offensive, Roman envoys presented the *fetiales* ritual, demanding restitution or surrender. Only when these demands were refused could war be declared as morally and legally justified. Yet Rome was also pragmatic, often using this ritualised restraint as a veneer over calculated aggression. Their legions perfected the art of systematic destruction—roads, siege engines, and formations that allowed annihilation on an industrial scale. The paradox was stark: the empire that gave us the idea of lawful war also razed Carthage to the ground, sowing salt into its soil. *Restraint and excess, principle and pragmatism, were not contradictions but coexisting elements of Roman statecraft.*

The Hebrew Bible presents another layer of complexity. In its narratives, war oscillates between divine command and ethical hesitation. The destruction of Jericho, with its total annihilation of inhabitants, reflects an uncompromising theology of war as purification. Yet prophetic voices, from Isaiah to Amos, criticised excessive violence and warned that unchecked destruction would invite divine judgement. This tension between command and conscience shows an early recognition that violence, once engineered and unleashed, could corrode not only enemies but the moral fabric of the victors themselves.

In India, ancient texts such as the *Mahābhārata* confronted these dilemmas with striking sophistication. The epic's central war is presented as both inevitable and tragic, and the *Bhagavad Gītā* stages its most famous philosophical dialogue on a battlefield. Arjuna, paralysed by the prospect of killing kin, receives instruction from Krishna that duty (*dharma*) may compel action even when destructive. Yet the text also places restrictions: rules of war prohibited fighting at night, harming unarmed soldiers, or attacking civilians.

Here, restraint was not an afterthought but an integral part of warfare's moral structure. The acknowledgement that power must be bound by principle demonstrates a recognition that technical capacity for destruction, left unchecked, could unravel cosmic and social order alike.

Chinese thought approached the ethics of destruction through a lens of harmony and pragmatism. Sun Tzu's *Art of War* famously valorises victory without battle, suggesting that the highest skill lies in subduing the enemy without wholesale destruction. This was not a pacifist sentiment but a practical recognition that total annihilation was often wasteful and destabilising. Confucian traditions reinforced the idea that war, while sometimes necessary, must serve the higher aim of restoring order. By contrast, Legalist thinkers saw no moral issue in ruthless methods if they preserved the state. The coexistence of these schools highlights the enduring tension between expediency and ethics, between engineering destruction for immediate gain and restraining it for long-term stability.

The technologies of destruction themselves posed ethical questions. The development of the catapult, for instance, made mass killing at a distance possible in ways that unsettled traditional notions of courage and honour. To kill without direct confrontation was, for some, a form of cowardice; for others, it was simply intelligent adaptation. Similarly, the burning of ships or crops as strategies of denial—scorched earth tactics—provoked debate about whether such acts were legitimate warfare or immoral devastation of the innocent. *Technological innovation always carried an ethical shadow, forcing societies to ask whether capability equated to legitimacy.*

What emerges from these diverse traditions is that ancient societies never saw destruction as purely technical. It was always bound up with stories, rituals, and philosophies that sought to legitimise or restrain its use. The act of building more efficient weapons or fortifications was paralleled by efforts to create moral frameworks that made sense of their deployment. Sometimes these frameworks were cynical masks for raw power; at other times, they represented genuine attempts to humanise violence. But in every case, the engineering of destruction demanded an accompanying engineering of justification.

This historical perspective resonates with contemporary debates. Modern warfare, with its drones, cyberweapons, and nuclear arsenals, continues to raise the same question: does the ability to destroy confer the right to do so? Ancient precedents remind us that the search for restraint is as old as the search for power. Just as the Romans cloaked conquest in ritual, or the Indians codified battlefield ethics, so too do modern states draft treaties, conventions, and doctrines. The tools change, but the intellectual struggle remains constant.

The ethics of destruction in antiquity reveal a profound truth: human beings have always sought to reconcile their ingenuity with their conscience. The capacity to engineer violence is matched by the capacity to reflect upon it, to ask whether survival, order, or glory is worth the cost in blood and ruin. These debates were not side notes to ancient history; they were central to how societies understood themselves. *The fortress, the bow, the catapult—all were inventions of the hand. But the decision to restrain or unleash them was an invention of the mind, a recognition that power without principle is itself a kind of defeat.*

Chapter 6
The Algorithm Before Numbers

Long before ink stained parchment or symbols were etched into clay, humans were already devising methods to organise thought, preserve memory, and execute sequences that we might now recognise as computational. The drive to impose order upon the world predates formal mathematics, predates written numerals even. A notch carved into bone, a cord knotted in a deliberate pattern, a rhythm tapped upon wood—these were more than mere marks or sounds. They were gestures in a choreography of abstraction, tools for compressing information into forms that could be stored, communicated, and repeated. In essence, what we now call algorithms arose from the human impulse to transform the ephemeral into the enduring.

What strikes one most about these early cognitive tools is their universality. Palaeolithic Europe has yielded tally sticks marked at regular intervals, suggesting not only counting but the recognition of cycles—perhaps lunar phases, perhaps hunting seasons. Across the Andes, the Inca devised the quipu: cords knotted and coloured in precise configurations to record censuses, tributes, and calendrical data with remarkable accuracy. Neither system relied on numerals as we understand them, yet both fulfilled the essential purpose of numbers: to organise quantity and sequence reliably. The significance lies not in material—bone, wood, or fibre—but in the abstract leap: thought encoded into a manipulable structure. To read a quipu was to engage with proto-code, a set of instructions intelligible to the trained eye.

Abstraction extended far beyond counting. Consider weaving: the patterns governing warp and weft were never purely decorative. The rules imposed an algorithmic discipline—follow the sequence and a repeatable, often meaningful, design emerged. Cloth became a medium of social identity, ritual, and memory. The parallels to modern information storage are uncanny: both rely on embedding meaning within reproducible structures. Music, especially in oral traditions, demanded the same precision. Rhythms passed down through generations were not merely aesthetic but served as methods for transmitting structure across time without writing.

Architecture too manifests early mathematical reasoning. The proportions of Egyptian temples, Greek sanctuaries, Mesoamerican pyramids, and Islamic cities reveal deliberate engagement with patterns that we might today describe as equations rendered in space. Builders, often illiterate in numerals, intuitively grasped symmetry, ratio, and proportion—concepts we teach in classrooms as mathematical fundamentals. An arch stands because its stones obey principles that transcend notation; mathematics exists in embodied form long before it is codified.

The later emergence of number systems extended these intuitive structures into philosophical territory. Mesopotamia's sexagesimal system enabled astronomers to model celestial cycles with startling accuracy, implicitly grappling with infinity. In India, the conceptual invention of zero transformed arithmetic from mere counting into a language capable of expressing absence, potential, and recursion. Zero was more than a placeholder; it was a philosophical declaration: nothingness could be quantified, could carry logical weight equal to something. Such concepts did not arise from idle speculation; they addressed concrete problems. How to track seasons, debts, or eclipses without a symbol for zero? How to reckon the boundless sky without conceiving the infinite? Practical necessity pushed thought toward abstraction, and abstraction reshaped the mind.

Commerce was a formidable engine of this development. Merchants in the ancient Near East recorded transactions not by tallying individual animals, but by categorising goods, allowing arithmetic reasoning to operate on classes rather than singular items. The Roman abacus employed beads representing units and tens, externalising procedure so that mental calculation could be performed physically. Here, tool and algorithm converge: moving beads according to rules enacted logic, materialising method.

Astronomy offered another crucible. Observing the heavens demanded recognition of patterns over vast temporal scales. Babylonian astronomers compiled diaries spanning centuries, understanding that prediction required nesting cycles within cycles. Their tables of planetary motion were methodical computations long before algebra existed. To forecast an eclipse centuries ahead, without calculus, illustrates the sophistication of pattern recognition in the absence of formal mathematics.

Across these examples, a unifying principle emerges: structure can substitute for symbols. Knot, pattern, rhythm, building—all are forms of externalised reasoning. Computation requires no silicon, no written digits, only the systematic application of rules to information, whether stored in fibre, stone, or sky. Ancient societies did not merely solve complex problems; they invented symbolic systems carrying logical weight before numbers themselves existed. The algorithm, it seems, predates arithmetic.

For the modern reader, this is no mere historical curiosity. Algorithms dominate our world today—from finance to social media—but their lineage stretches back to shepherds scratching marks on bone, to priests tying knots in quipu. The ancient world demonstrates that algorithms are not bound to numerals but to the human capacity for pattern recognition, abstraction, and structured procedure. Understanding this pre-numerical ingenuity broadens our sense of mathematics: it is not merely symbols on paper but strategies for ordering reality.

This chapter traces that spectrum. It begins with counting without digits, exploring how tallies and quipu functioned as abstract computational devices. It moves through the encoding of reasoning in patterns—woven cloth, music, architecture—where structure carried logic. It then examines the intellectual revolutions of zero and infinity, confronting absence and boundlessness. Finally, it surveys the practical mathematics of merchants, builders, and astronomers, whose daily challenges embedded calculation into life itself. What emerges is a portrait of mathematics not as a late invention but as an ancient human instinct, expressed long before written numbers formalised it.

To follow this history is to recognise a continuity: our algorithms, however complex, continue the project begun millennia ago—to tame uncertainty by encoding thought into repeatable procedures. Whether inscribed in clay, knotted in cord, or written in code, the principle remains unchanged: to make the intangible graspable, the fleeting durable, the infinite manageable. The algorithm before numbers is the story of how humans first learned to think with structure—and in doing so, laid the foundations for all subsequent mathematics.

Counting Without Digits

The first humans who sought to measure the world did not begin with numbers but with marks. A notched bone from the Upper Palaeolithic, dating back some 30,000 years, bears a sequence of incisions whose regularity suggests more than idle scratching. Some interpret it as a lunar calendar, others as a tally of prey or days. Whatever its specific function, it reveals a critical truth: *counting preceded numbers*. Long before abstract symbols like "5" or "10" came into being, people devised systems for recording and manipulating quantity, externalising memory into material form. These were not crude preludes to mathematics but sophisticated strategies of abstraction, capable of storing, transmitting, and processing information in ways that rival the power of written numerals.

The tally stick is among the most enduring of these devices. Found across Europe, Africa, and Asia, it consisted of a piece of wood or bone with incised marks, each signifying an item, day, or event. The famous Ishango bone, unearthed near the headwaters of the Nile and dated to around 20,000 years ago, bears a series of parallel notches arranged in groups that some scholars read as arithmetic operations, possibly even evidence of prime number recognition. Whether or not the makers consciously grasped such concepts, the artefact demonstrates the capacity to embed systematic thought into marks. A tally stick is not simply a reminder—it is a method, a way of delegating cognitive load to the material world. One could add a notch instead of recalling a running total, turning arithmetic into a physical process. In this sense, *a tally stick is both record and algorithm: it instructs the user on how to continue the count.*

What made tally systems powerful was their scalability and social application. In medieval Europe, split tally sticks became a trusted means of financial accounting. A piece of hazelwood would be notched to denote a debt, then split lengthwise: one half retained by the creditor, the other by the debtor. Because the grain of the wood matched uniquely, forgery was nearly impossible, and mutual trust could be formalised without literacy or coinage. Here the tally became more than memory; it was contract, a binding instrument of exchange. Its authority derived not from numerical literacy but from material integrity. In this way, tally sticks illustrate how societies solved complex problems of record-keeping and trust long before standardised numerals or modern financial systems.

The Inca quipu represents an even more intricate leap in pre-numerical computation. Made of coloured cords, each bearing knots tied in specific positions, the quipu could encode numerical data through place value. Tens, hundreds, and thousands were represented not with written digits but through the position and type of knot along a string. More remarkably, colour and cord arrangement allowed for layers of meaning beyond mere quantities: census records, tribute obligations, and possibly even narratives may have been preserved. Spanish chroniclers after the conquest noted that quipu-keepers, or *khipukamayuq*, could read these cords with a fluency comparable to literacy, extracting demographic, economic, and historical information. The quipu thus challenges our assumptions about what counts as writing or computation.

If data can be stored, retrieved, and interpreted systematically, does it matter whether the medium is cord, clay, or binary code?

Understanding quipu as an abstract computational tool requires seeing beyond its physical form. Each knot was not just a representation of a number but an instruction in a positional system, requiring consistent rules for interpretation. This made quipu capable of operations such as addition, subtraction, and possibly multiplication. A census recorded in quipu format could be updated by retying or adding knots, enabling dynamic data management. In this sense, the quipu functions analogously to a modern database: structured, extensible, and interpretable by trained users. While not digital, it embodies many principles that underpin computation—representation, positionality, and repeatable operations.

What unites tally sticks and quipu is not simply their role as counting aids but their function as externalised cognition. To use them was to transfer mental processes into physical artefacts, reducing the burden of memory and allowing complexity to scale. This is the essence of abstraction: by encoding a system into a material device, human societies amplified their capacity to think, organise, and plan. One might say that before numbers, people already had algorithms—sequences of steps embedded in material form, guiding the user from input (a sheep passing a gate, a tax paid) to output (a notch, a knot). The invention of numerals refined this process, but it did not create it.

The philosophical implications are significant. We often imagine mathematics as beginning with written symbols, but the history of tally systems and knot records shows that *computation is older than notation.* The human drive to quantify, structure, and codify experience found expression not in abstract digits but in embodied practices—marks on bone, cords of fibre, sequences of knots. These systems remind us that the essence of mathematics is not in the signs themselves but in the relationships they encode. A knot tied in the right place carries as much logical weight as the numeral "7."

The resonance with modern technologies is striking. Just as a tally stick externalised memory into wood, today's spreadsheets externalise calculation into software. Both are tools for offloading cognition, ensuring accuracy, and enabling scale. The quipu, in particular, anticipates the logic of positional notation and database design, demonstrating that the distinction between "primitive" and "modern" computation is one of form, not function. When we reduce algorithms to digits on screens, we risk overlooking the ingenuity of those who computed without numbers, building robust systems of abstraction from the simplest of materials.

Ultimately, counting without digits was not a failure of imagination but a triumph of it. To mark, knot, or arrange is to assert that thought can be structured and externalised, that information can be made durable and transferable. The tally and the quipu stand as evidence that the algorithmic mind predates the numerical one, and that human beings, from the earliest times, sought not only to count the world but to encode it. They remind us that mathematics is not merely about symbols; it is about the human impulse to create systems that make reality comprehensible and navigable. And in that respect, the cord and the notch stand in direct lineage with the equation and the code.

Patterns as Equations

Before symbols were written on clay or parchment, mathematics was already woven into the fabric of human activity—quite literally. The earliest systematic reasoning about numbers and proportions did not take the form of abstract formulae but of patterns: warp and weft crossing in deliberate sequences, notes strung in intervals that pleased the ear, stones laid with geometric precision to hold against gravity. To see weaving, music, and architecture as aesthetic crafts alone is to underestimate them. They were, in fact, living laboratories of mathematical reasoning, encoding relationships of proportion, repetition, and transformation long before algebra gave such relationships their formal notation.

Consider weaving. A loom is more than a tool for producing cloth; it is a machine for executing algorithms. The alternation of warp and weft threads, governed by strict rules of over and under, creates patterns that can be extended indefinitely. A simple alternation generates plain weave, while more complex sequences yield twills, herringbones, or brocades. These are not merely decorative. Each represents a systematic manipulation of binary choices, a kind of embodied logic. The weaver must hold in mind the sequence of lifts and passes, translating abstract pattern into material outcome. To design a new weave is to conceptualise structure in terms of repetition and variation—an act indistinguishable, in essence, from constructing an equation. Indeed, modern computer scientists studying weaving have noted its kinship with binary coding: every thread position is either raised or lowered, a one or a zero. What looks like fabric is in fact a highly compressed record of mathematical operations.

Anthropologists studying traditional textiles from cultures as distant as the Navajo of North America, the Dogon of West Africa, and the Andean peoples have found similar systems of encoded complexity. Certain weaves required recursive sequences—where the end of one pattern cycle initiates the next—mirroring the logic of iterative algorithms. Others exploited symmetry and mirroring, creating designs that obey the principles of group theory centuries before mathematicians defined it. The weaver, of course, did not name these concepts, but the fabric in her hands was a tangible manifestation of their truth. *Patterns here were not ornamental distractions from survival but instruments of cognition, encoding rules in forms that could be seen, touched, and replicated.*

Music, too, operates as mathematics made audible. The Pythagoreans famously measured consonant intervals in terms of string lengths, discovering that the octave, fifth, and fourth could be expressed as simple ratios—2:1, 3:2, 4:3. This was more than a curiosity about harmony. It revealed that pleasing sound was structured by numerical relationships, and that patterns of vibration could be predicted, controlled, and systematised. The discipline of music theory emerged not as an art of expression alone but as a science of proportion. When Indian theorists mapped ragas or when medieval monks codified Gregorian chant, they were not simply prescribing aesthetic rules but formalising sequences that obeyed mathematical reasoning. Rhythm, too, is a study in ratios. A drum pattern alternating two beats and three creates a cycle of five, which when repeated against a four-beat metre generates polyrhythm—

a practical demonstration of least common multiples. To perform such rhythms is to embody modular arithmetic.

The architectural tradition offers perhaps the most visible legacy of patterns as equations. From the Egyptian pyramids to the Gothic cathedrals of Europe, builders manipulated geometry not for abstraction's sake but to create stable, enduring structures that also satisfied aesthetic ideals. The use of proportional canons—such as the "golden ratio" in Greek temples or the harmonic ratios encoded in Islamic design—was not arbitrary ornamentation. These proportions were equations expressed in stone, harmonising visual balance with structural necessity. The pointed arch of the Gothic style, for instance, is not merely decorative but a solution to distributing vertical load into lateral thrusts, encoded in the geometry of intersecting circles. To construct such an arch required an intuitive grasp of geometric principles, translated into rules of practice long before Euclid's axioms were formalised.

What unites weaving, music, and architecture is their ability to make abstract relations tangible. Each domain embeds logic into form: the repetition of threads, the sequence of tones, the alignment of stones. In each, the practitioner engages in problem-solving that requires anticipation, adjustment, and systemic thinking. Crucially, these activities were accessible without written numerals. A pattern could be memorised, taught, and transmitted through demonstration. The loom, the lyre, and the compass became pedagogical devices for entire communities, carrying mathematical reasoning in embodied, collective practice. *Equations were not first written on paper; they were first enacted in rhythm, pattern, and proportion.*

One might argue that these encoded systems were proto-mathematics, awaiting the refinement of symbols to achieve their full potential. But this perspective undervalues their autonomy. The quipu of the Andes or the rhythmic cycles of Indian tala are not failed precursors of modern algebra or calculus; they are complete systems in themselves, tailored to the needs and perceptions of their cultures. To analyse them through the lens of Western mathematics risks anachronism. Better to see them as alternative expressions of the human capacity to formalise thought—parallel routes to the same impulse to capture regularity, to abstract from the flux of experience a repeatable order.

The modern sciences continue to rediscover the power of patterns. In computer science, algorithms are often taught through visual patterns such as sorting diagrams, echoing the loom's sequences. In physics, the search for symmetry remains the guiding principle of theory-building, from crystallography to string theory. In data analysis, patterns are sought as signals amidst noise, revealing correlations otherwise invisible. These endeavours differ in form but not in kind from the weaver's recursive motifs or the musician's rhythmic cycles. *To recognise pattern is to glimpse structure beneath appearance, and to encode pattern is to wield that structure as a tool.*

It is tempting to think of equations as inventions of scribes, locked in the notational conventions of mathematics. Yet equations existed wherever human beings confronted complexity and sought order. A woven textile, a harmonic scale, or a vaulted ceiling all testify to the same truth: that long before symbols, humans thought in patterns, and those patterns *were* their equations. To study them today is not to romanticise the past but to acknowledge that abstraction has many languages. Stone, string, and sound have been, in their own right, media of calculation. To treat them as crafts without intellectual substance is to overlook their role in shaping the very idea of reasoning itself.

Zero and Infinity

Few intellectual leaps in human history rival the audacity of inventing zero or conceiving infinity. Both concepts appear deceptively simple to the modern mind—schoolchildren casually subtract with zero, mathematicians freely manipulate infinite series—but their arrival in human thought was neither natural nor inevitable. For centuries, civilisations managed commerce, architecture, and astronomy without needing to define either nothingness or boundlessness in numerical terms. Yet in India and Mesopotamia, these abstractions emerged not only as practical tools but as profound philosophical gestures. They signalled a willingness to grapple with the void and with the limitless—ideas that stretch the limits of reason itself.

Mesopotamia offers one of the earliest recorded steps towards zero. By the third millennium BCE, the Babylonians had developed a positional number system based on base sixty, a remarkable advance that allowed large numbers to be represented with compact efficiency. Yet such a system quickly runs into a problem: how does one mark an empty place? Without a sign to distinguish, for instance, between 61 and 3601, positional notation collapses into ambiguity. The Babylonians solved this pragmatically by introducing a placeholder symbol—a pair of angled wedges—to indicate absence within a numeral. This was not yet zero as a number in its own right but rather a gap-marker, a sign that absence had to be accounted for if calculation was to remain precise. Their astronomical tables, some of the most sophisticated computational documents of the ancient world, depended on this placeholder to calculate planetary movements across vast timescales. *Absence here became not an afterthought but a structural necessity of reasoning.*

India carried the concept further, transforming absence into an entity. By the fifth century CE, Indian mathematicians had begun to treat zero not merely as an empty placeholder but as a number with its own properties. The mathematician Brahmagupta, writing in his *Brahmasphutasiddhanta* (628 CE), laid down rules for arithmetic operations involving zero: a number minus itself is zero, zero plus zero is zero, and intriguingly, zero divided by zero is undefined. What distinguished this leap was not only its technical ingenuity but its cultural resonance. Indian philosophy had long entertained the notion of *śūnya*—emptiness or void—as a metaphysical principle, particularly within Buddhist and Hindu traditions. The mathematical zero thus resonated with a worldview that did not fear nothingness but understood it as foundational. Zero, once grasped, did not remain confined to mathematics. It travelled along trade and intellectual routes into the Islamic world, where it was absorbed into algebra, and from there into Europe, reshaping science and commerce alike.

Infinity presented an even greater conceptual challenge. Whereas zero asked whether nothing could be something, infinity asked whether something could ever be endless. Mesopotamian thought treated vastness pragmatically, as in astronomical cycles that extended into enormous numbers, but rarely entertained infinity as a distinct entity. Greek philosophy, by contrast, wrestled directly with the idea. Aristotle distinguished between potential infinity—processes that could continue without end, such as dividing a line into ever smaller parts—and actual infinity, an existing infinite totality, which he rejected as incoherent. For centuries, this distinction haunted Western thought, keeping infinity at the edges of acceptable reasoning.

India, however, once again displayed a remarkable openness. Jain mathematics, developing from the first millennium BCE, elaborated a taxonomy of infinities: some sets were infinite in one direction, others in two; some were infinitely divisible, others infinitely extensible. This was not mystical speculation but systematic categorisation, motivated by theological and cosmological debates about the nature of time, space, and the soul. Such thinking seeded a culture in which infinity could be discussed as a legitimate mathematical object. Later, texts such as the *Surya Siddhanta* incorporated infinite processes into astronomical calculation, anticipating techniques that would not reappear in Europe until the calculus of Newton and Leibniz.

What makes zero and infinity extraordinary is not merely their mathematical utility but the *intellectual courage* they demanded. Both concepts require moving beyond the tangible. Merchants could count goods, builders could measure stones, but how does one grasp nothing? How does one calculate with the endless? To accept zero was to accept that absence could function like presence, that the void could be part of structure. To accept infinity was to admit that human reason could represent what it could never fully traverse. Each concept demanded an act of imagination: the leap from counting apples to contemplating the infinite division of a line, from recording debts to recognising the generative power of nothingness.

The resistance these ideas faced underscores their radicality. Roman numerals had no zero, and the Roman world managed without it for centuries, relying on concrete counting methods. Medieval Europe initially resisted the "Arabic numerals" imported from India via the Islamic world, suspecting them as foreign and even heretical. Infinity fared no better: European scholastics debated whether God alone was infinite, and Galileo's paradoxes of infinity unsettled the foundations of mathematics well into the seventeenth century. These were not trivial controversies but clashes at the frontier of what human thought dared to formalise.

The modern world rests upon these once-fragile abstractions. Digital technology, for instance, depends on binary code, where the distinction between one and zero drives all computation. Zero here is not passive absence but active state, shaping every logic gate and algorithm. Infinity, meanwhile, underpins modern physics and cosmology: the expansion of the universe, the notion of infinite divisibility at quantum scales, the infinite set theory developed by Georg Cantor in the nineteenth century. Yet it is striking to remember that these modern uses extend ideas first forged by ancient minds confronting the metaphysical edges of reality.

To appreciate the origins of zero and infinity is to see how mathematics is inseparable from philosophy. These were not mere technical conveniences but *philosophical positions embodied in symbols*. They forced ancient thinkers to confront the boundaries of language, logic, and metaphysics, producing insights that continue to reverberate. If Mesopotamia's wedge-shaped placeholder revealed the necessity of marking absence, and India's śūnya elevated that absence into presence, the conversation across cultures reveals the multiplicity of ways humans sought to think the unthinkable. If Jainism could imagine categories of infinity and Aristotle could debate its impossibility, both demonstrate that mathematics is as much about what we are willing to conceptualise as what we can measure.

In recognising this, we glimpse why zero and infinity remain not just mathematical tools but philosophical provocations. They remind us that abstraction is not sterile; it is daring. To invent them was to stretch thought into domains where intuition falters, to risk paradox for the sake of coherence. That ancient societies could make these leaps without modern formalism testifies to the extraordinary reach of human imagination. We owe to them not only our ability to calculate interest rates or engineer spacecraft but our very capacity to *think beyond the limits of the visible*.

Practical Mathematics

The idea that mathematics once lived only in dusty scrolls or the minds of philosophers is a distortion born of hindsight. In antiquity, mathematics was not an abstract discipline divorced from life; it was woven into the daily practices of trade, construction, and the observation of the heavens. Merchants weighed, measured, and calculated; builders applied geometry to raise temples and aqueducts; astronomers relied on numerical models to anticipate the movements of celestial bodies that governed agriculture and ritual. What we call "practical mathematics" was not a diluted version of theoretical knowledge but its proving ground, a domain where the necessity of solving real problems drove innovation, precision, and conceptual breakthroughs that later generations would codify as theory. *It was in the marketplace, the quarry, and the night sky that the abstractions of mathematics earned their relevance.*

The merchant was one of the first practitioners of mathematics in a structured sense. Long before algebraic symbols existed, traders operating along the Nile or the Euphrates had to maintain accurate accounts of goods exchanged, debts owed, and payments expected. Clay tablets from Mesopotamia, dating as far back as 2000 BCE, record complex commercial transactions involving interest rates, deferred payments, and commodity prices. These are not simplistic tallies but structured contracts that required a command of proportional reasoning and compound calculations. The Babylonian sexagesimal system—based on the number sixty—emerged from this world, not because sixty had mystical significance but because its divisibility made fractions manageable. A merchant calculating one-third of a shipment of grain or one-fifteenth of a herd of sheep could express the outcome with clarity in base sixty in a way base ten could not easily accommodate. Here, practicality dictated structure.

The Egyptian economy shows a similar pattern. Papyrus documents such as the Rhind Mathematical Papyrus (c. 1650 BCE) preserve problem sets that at first glance look like school exercises. Yet the problems themselves reveal pressing commercial concerns: how to distribute bread and beer fairly among workers, how to divide land after the annual Nile flood, how to measure grain storage. The solutions employ unit fractions—breaking quantities into reciprocal parts—which may seem cumbersome to modern eyes but provided an elegant way to ensure exact division in a society obsessed with fairness of allocation. The fact that such "exercises" were copied and taught indicates that these were not idle puzzles but tools of economic survival. *The abstract formula emerged only after the real-life question demanded an answer.*

Building projects brought another layer of mathematical rigour. Consider the construction of the Egyptian pyramids: massive stones cut with remarkable precision, transported over long distances, and arranged into structures whose alignment with cardinal points remains awe-inspiring. Geometry here was not an intellectual indulgence but a logistical necessity. To maintain the pyramid's slope, surveyors employed right triangles formed by ropes knotted at regular intervals, creating a reliable template for angles. The concept of a "rope-stretcher"—a surveyor who laid out fields or construction lines—appears in Egyptian records and survives as a reminder that geometry, from its very etymology (*geo* meaning earth, *metron* meaning measure), was born from the land and its manipulation. Mesopotamian builders, too, deployed precise measurements in constructing ziggurats, the tiered temples that dominated city skylines. Their bricks bore standardised dimensions, and surviving tablets suggest ratios were carefully maintained to ensure stability. Architecture provided the crucible where arithmetic encountered proportion, symmetry, and the beginnings of trigonometric reasoning.

Astronomy elevated mathematics to a domain that merged practicality with transcendence. Agricultural societies depended on accurate calendars, and astronomy provided the predictive framework to anticipate planting and harvest seasons. Babylonian astronomers, building on centuries of observation, developed numerical schemes that could forecast eclipses and planetary motions. Their methods were astonishingly algorithmic: they recorded data, detected periodicities, and extrapolated forward. What they lacked in geometric models they compensated for with tables of numbers that effectively captured celestial cycles. This was mathematics serving both priestly ritual and economic planning, since knowing when the floods or harvests would come was as much about survival as ceremony. In Egypt, the heliacal rising of Sirius became the anchor of the calendar, ensuring that farming cycles aligned with the life-giving inundation of the Nile.

The Greek contribution often appears framed as a pivot from practicality to theory, but this narrative conceals continuity. The Greeks certainly elevated mathematics into an abstract science—Euclid's *Elements* is paradigmatic of deductive reasoning—but even here, practical needs underpinned development. Greek sailors navigating the Mediterranean required trigonometric tools, and builders of amphitheatres or aqueducts grappled with acoustics and hydraulics, both deeply mathematical challenges. The Pythagorean theorem, immortalised as a triumph of abstract reasoning, was known in practice to Egyptian and Babylonian surveyors long before it was proved deductively. Its proof may have been Greek, but its use was ancient and pragmatic. *Theory did not precede practice; it codified and refined it.*

One of the most illuminating case studies lies in the way interest and debt were managed in Mesopotamia. Tablets show detailed calculations of compound interest over long periods, with scribes applying consistent formulas centuries before the concept was articulated in European mathematics. A loan of silver at 20 per cent annual interest, rolled over for several years, demanded exponential thinking, and the scribes executed it with fluency. That such mathematics arose from commerce rather than contemplation challenges the modern stereotype that finance corrupts intellectual purity. On the contrary, commerce acted as a relentless laboratory, forcing clarity where imprecision carried ruinous costs. The need to settle accounts created the pressure to invent and standardise techniques that endured across civilisations.

A tension emerges here between mathematics as a servant of necessity and mathematics as an instrument of intellectual exploration. Merchants wanted efficient calculation; astronomers sought predictability; builders demanded stability. But each time mathematics delivered a solution, it also opened a question: if fractions divide evenly here, why not generalise? If planetary cycles repeat, what governs their recurrence? If triangles produce consistent ratios, what deeper principles underlie them? The Greeks gave voice to these questions with philosophical rigour, but the spark was kindled earlier in the sweat and dust of daily problem-solving. *The abstract and the practical are not separate strands but interwoven threads of one fabric.*

Practical mathematics also exposes a cultural dimension often neglected in modern narratives. The Indian *Sulba Sutras* (c. 800–200 BCE), ritual texts on altar construction, contain strikingly precise geometric rules. The instructions describe how to transform one shape into another of equal area, ensuring ritual altars satisfied cosmological requirements. Here mathematics was neither purely commercial nor purely scientific but religious, embedded in the structure of spiritual life. Similarly, in China, the *Nine Chapters on the Mathematical Art* (compiled around the first century CE) included problems on taxation, engineering, and surveying, illustrating that mathematics was inseparable from governance and infrastructure. Across these traditions, practical necessity and cultural meaning fused into systems of knowledge that were anything but rudimentary.

Modern readers might assume that with our electronic calculators and sophisticated algorithms, we have left such rudimentary practicalities behind. Yet every swipe of a credit card or global positioning signal echoes the same logic. Financial mathematics remains at the heart of commerce, though now obscured by abstraction; civil engineering still relies on principles not wholly different from those that guided pyramid builders; astronomy has been expanded into astrophysics, yet its core is still the numerical prediction of celestial motion. The everyday mathematics of antiquity survives in our institutions, our technologies, and even our habits of thought.

What is most striking is how these early forms of mathematics preserved a humility before experience. Merchants did not calculate to demonstrate intellectual brilliance but to ensure survival and fairness. Builders measured angles because miscalculation risked collapse. Astronomers charted the skies not to speculate idly but to secure harvests and rituals. In each case, the discipline of mathematics was anchored to the practical, and this grounding protected it from the sterile excesses of abstraction detached from consequence. The lesson is enduring: when mathematics stays close to lived reality, it flourishes as a tool of empowerment rather than exclusion.

Thus, the story of practical mathematics is not that of a "primitive precursor" to modern theory but of a dynamic, adaptive intelligence responding to the demands of survival, justice, and meaning. From the trader's clay tablet to the mason's rope to the astronomer's star chart, mathematics was lived and enacted. Its authority rested not in ivory towers but in granaries, temples, and fields. If we look closely, the abstractions we now prize—the algebra, the geometry, the trigonometry—were crystallised out of these practical engagements. They remind us that mathematics, at its heart, is not only about numbers and proofs but about human ingenuity at work in the world.

Chapter 7
Medicine Without Molecules

Long before microscopes revealed microbes or laboratories dissected the secrets of DNA, human societies confronted illness with observation, analogy, ritual, and craft. For the vast expanse of our history, medicine did not rely on chemical compounds or test tubes but on frameworks that translated experience into action. Healing was less an application of molecular knowledge than a structured engagement with patterns—of the body, of the environment, and of the mind. To dismiss these traditions as superstition would be to overlook the sophistication of their logic, their resilience across centuries, and the intellectual scaffolding that allowed communities to navigate disease long before cells or germs had names.

Analogy was central to early medical thought. In Greece, physicians interpreted health through humours: blood, phlegm, black bile, and yellow bile. Disease, they believed, arose from imbalance, and remedies—dietary adjustments, bloodletting, herbal preparations—were designed to restore harmony. The system may seem foreign today, yet it provided an internally coherent logic and a shared language for diagnosis and treatment. Across China, a similar commitment to relational thinking appeared in the yin-yang framework, where health depended on balance and qi's unobstructed flow. Disruption of these forces demanded interventions ranging from acupuncture to herbal therapy. Such models may not map onto modern biology, but they represent disciplined attempts to render illness intelligible.

Elsewhere, Ayurveda in India offered a comparable vision through the doshas—vata, pitta, and kapha—three fundamental forces shaping body, mind, and environment. Its insight lay less in anatomical precision than in integrating lived experience with preventative and curative practice. These systems embedded medical reasoning into daily life, guiding diet, exercise, and lifestyle, and they persisted across generations precisely because they offered a stable framework in the absence of diagnostic imaging or pharmaceuticals.

Alongside conceptual models, practical intervention was remarkably advanced. Archaeological evidence attests to early surgery: trepanation—drilling or scraping into the skull—has been documented across Neolithic Europe, Africa, and South America, with many patients surviving the procedure. Whether intended to release "evil spirits" or relieve cranial pressure, such operations reveal technical skill and a tradition of experimentation. Egyptian prosthetics, too—wooden toes designed for mobility rather than decoration—demonstrate resourceful adaptation anticipating modern design principles.

Herbal anaesthetics and antiseptics further complicate any simplistic notion of "primitive" medicine. Communities worldwide identified plants with sedative or numbing effects: coca in South America, opium and mandrake in the Mediterranean, cannabis in India and China. These discoveries arose from cumulative observation, trial, and careful practice. Without knowledge of alkaloids, early healers cultivated effective pharmacopoeias—some crude, some surprisingly potent—that modern chemistry would later validate.

Medicine was inseparable from ritual and spirituality. Healing addressed not only the body but also social and cosmic order. In Mesopotamia, sickness might signify divine displeasure, treated with incantations alongside physical remedies. Greek temples of Asclepius combined purification, sleep therapy, and priestly consultation. In China, medical practice aligned with cosmology, and among the Maya, herbs, prayers, and offerings were interwoven. These practices underscore that medicine aimed to restore equilibrium—social, spiritual, and corporeal alike.

Ritual, far from irrelevant, mobilised psychological and communal resources. Placebo effects remind us that belief, expectation, and cultural meaning materially affect recovery. A patient comforted by ritual might sleep more, eat better, and reduce stress—all tangible contributors to healing. Ancient medicine demonstrates a prescient understanding of the mind-body interplay that contemporary research is only now quantifying.

Trial and error underpinned all these systems. Remedies were retained or discarded based on observed outcomes, regardless of theoretical justification. Honey as a wound dressing, quinine bark for fevers—these interventions often preceded mechanistic explanation by centuries. Empirical practice advanced health care even absent molecular insight.

The philosophical import is clear: knowledge can accumulate and adapt without modern "scientific correctness." Humoural theories or yin-yang provided coherence; surgery pushed physical possibility; ritual mobilised psychosocial resources; empirical observation refined treatments. Together, these elements sustained human health across millennia.

The common thread is human refusal to surrender to illness. Across continents and centuries, communities devised symbolic, technical, and experimental strategies to confront suffering. These frameworks were not mere precursors to modern biomedicine but sophisticated responses calibrated to available knowledge and resources. Medicine, therefore, is not solely about molecules; it is an enterprise of culture, meaning, and ingenuity.

This chapter examines those intertwined dimensions. Readers will encounter practices that alleviated suffering and prolonged life without recourse to modern laboratories, revealing how early societies constructed knowledge, refined techniques, and embedded healing within broader social and cosmological systems. Their legacy is not only in remedies preserved but in a profound understanding: progress is rarely linear, meaning shapes outcomes as much as mechanism, and healing has always been as much art and philosophy as science.

Diagnosing by Analogy

When ancient physicians sought to understand why the body faltered, they did not peer through microscopes or measure blood chemistry. Instead, they turned to analogy: systems of thought that mapped illness onto patterns of balance, imbalance, flow, and obstruction. To the modern mind, accustomed to pinpointing disease through bacterial cultures or MRI scans, such methods may appear alien or imprecise. Yet these analogical frameworks were not haphazard inventions. They were intellectual architectures that made sense of suffering in the absence of molecular science, allowing healers to diagnose, predict, and treat with remarkable coherence. Their endurance across cultures and millennia attests not to superstition but to the persuasive power of metaphor in structuring human thought.

In the Greek world, medicine developed under the influence of the humoral theory, a model that compared the body to a delicate system of fluids. Blood, phlegm, yellow bile, and black bile were thought to govern health, with illness arising when one dominated the others. A fever was not caused by pathogens but by an overabundance of blood; melancholy, by an excess of black bile. While modern science recognises no such substances, the humoral framework offered a rational method of interpretation. Symptoms were read as signs of imbalance, and treatments—dietary adjustments, bloodletting, purgatives—sought to restore equilibrium. This was not wild guesswork but a logical system rooted in the observation of recurrent patterns. The model also dovetailed with broader Greek philosophies of harmony, proportion, and order. To speak of a body "out of balance" resonated not just medically but culturally, situating personal health within the cosmos' larger rhythm.

In China, another analogical model emerged, equally sweeping in scope yet distinct in its terms. Here, the human body was seen as an embodiment of yin and yang, complementary forces whose balance determined vitality. Illness was interpreted as the predominance of one force over the other, or as a disruption in the flow of *qi*, the life energy coursing through meridians. A cough might signal not merely lung irritation but an imbalance between dryness and moisture, between cold and heat. Diagnosis involved interpreting a constellation of signs—the colour of the tongue, the rhythm of the pulse, the sound of breathing—each linked metaphorically to natural forces. Treatment, whether through acupuncture, herbal decoctions, or lifestyle changes, sought not just symptom relief but a recalibration of these interwoven forces. Unlike humoralism, which emphasised internal fluids, Chinese medicine integrated the body into a broader cosmological system: health mirrored the balance of heaven and earth.

India's Ayurvedic tradition provides yet another case study in analogical medicine. Here, health was governed by the three doshas—vata, pitta, and kapha—each representing elemental forces of motion, transformation, and stability. Just as the humours and yin-yang corresponded to bodily functions, the doshas offered a lens through which symptoms could be interpreted. A person prone to anxiety and insomnia might be seen as vata-dominant, requiring grounding foods and calming routines. A person with inflammatory conditions might have aggravated pitta, treated with cooling herbs and avoidance of heat-producing behaviours. The system was comprehensive, spanning not only physical symptoms but psychological tendencies, lifestyle habits, and environmental influences. In doing so, it turned diagnosis into a form of narrative: the patient's life and body were read as expressions of imbalance, and healing was cast as a return to alignment.

What these traditions share is not a common set of substances or energies but a shared method of reasoning. *Analogy allowed healers to move from the visible to the invisible, from symptoms to causes, without recourse to unseen microbes or chemical assays.* This mode of thought has a certain elegance.

A fever is like fire; melancholy is like darkness; breathlessness is like blocked wind. These metaphors created a bridge between the abstract and the tangible, between the cosmos and the individual body. They provided coherence in a world where disease otherwise appeared arbitrary, offering explanations that were not only medically actionable but psychologically consoling.

Of course, analogical diagnosis was not without its limitations. Treatments based on humoral imbalance sometimes caused more harm than good, as in the overuse of bloodletting. The reliance on cosmic categories meant that empirical anomalies could be forced to fit preconceived frameworks, stifling alternative interpretations. Yet it is worth noting that modern medicine, too, employs analogy. Doctors still speak of "inflammation" spreading like fire, of "blocked arteries," of "viral invasions." The metaphors may be grounded in molecular reality, but they shape thought and guide intervention in ways strikingly continuous with earlier traditions. The difference lies not in the abandonment of analogy but in the refinement of its reference points.

The endurance of these analogical systems across centuries raises an important philosophical question: why did they persist so long despite their divergence from biological reality? One reason is practical efficacy. Many recommended treatments—dietary adjustments, herbal remedies, attention to lifestyle—did in fact alleviate symptoms, even if the explanations were wrong. Another reason is cultural integration. The humours echoed Greek ideas of harmony; yin and yang expressed Chinese cosmology; doshas embodied Indian metaphysics. Medicine was not a discrete field but part of a larger worldview, making its analogies compelling beyond the clinic. To challenge them would have required overturning entire cultural systems, not just medical theories.

It is also crucial to recognise the psychological and social function of analogical diagnosis. By situating illness within a framework of balance and imbalance, healers offered patients a sense of intelligibility and control. Disease was not random misfortune but a state that could be understood, managed, and reversed. This interpretive power should not be underestimated. Even today, patients often crave narratives that make sense of their suffering—why they fell ill, what it means, and how it can be corrected. Modern biomedicine, with its reduction to molecules, sometimes struggles to provide such narratives, leaving space for alternative systems that offer more holistic stories.

The study of ancient diagnosis by analogy therefore reveals more than a medical curiosity; it uncovers a deep structure of human cognition. Faced with uncertainty, we reach for metaphor. We make sense of the unknown by mapping it onto what is familiar, whether fluids, energies, or cosmic dualities. In this sense, ancient healers were not irrational but profoundly human in their reasoning. Their frameworks remind us that medicine is never just about mechanisms but also about meaning, and that patients need both to endure suffering and pursue recovery.

Understanding this legacy also sharpens our perception of modern medical practice. The precision of laboratory tests has not eliminated the need for interpretive frameworks; it has only shifted them. When doctors speak of "metabolic balance," "immune overreaction," or "hormonal regulation," they are, in effect, using new analogies to structure understanding. The metaphors differ, but the impulse to connect symptom and system through analogy remains unchanged. *What has evolved is the anchoring of these analogies in experimentally verifiable processes rather than in cosmology or philosophy.*

Diagnosing by analogy thus represents both a limitation and a strength in the history of medicine. It lacked the molecular accuracy of contemporary science, yet it offered coherence, narrative, and practical guidance that sustained communities for millennia. To dismiss it as mere superstition would be to miss its intellectual depth and cultural significance. At the same time, recognising its limitations prevents us from romanticising the past. The challenge for modern readers is to appreciate the ingenuity of ancient frameworks while understanding how they differ from our own.

By examining how healers once read bodies as texts of balance, flow, and correspondence, we gain insight into the enduring human need to render illness meaningful. And in recognising the continuities between ancient and modern metaphor, we may discover that the distance between humours, yin and yang, doshas, and today's "immune system out of balance" is not as great as it first appears. The language has changed, but the structure of thought endures.

Surgery and Experimentation

Few practices reveal the ingenuity and boldness of early medicine more vividly than surgery. Long before antiseptics, germ theory, or anaesthetic gases, ancient healers dared to cut into the body, replace lost parts, and dull pain with substances drawn from plants. These were not reckless gambles but calculated interventions carried out with tools of stone, bronze, or iron, guided by tradition, empirical learning, and a willingness to experiment. The very survival of patients—sometimes over decades—demonstrates that these early practitioners achieved more than mere ritual theatre; they succeeded in navigating the dangerous threshold between injury and healing in ways that demand recognition.

The earliest and perhaps most startling evidence of surgical intervention is trepanation: the act of drilling or scraping a hole into the skull. Archaeological discoveries reveal trepanned skulls dating back at least seven thousand years, found across continents from Peru to France to Central Asia. The reasons for the procedure varied—relieving pressure after head trauma, attempting to cure seizures or mental disturbances, or releasing perceived malevolent forces—but what stands out is not only the act itself but the survival it entailed. Many skulls show clear signs of bone regrowth, indicating that patients lived long after the operation. To perform cranial surgery with stone tools, without sterilisation or modern anaesthesia, and still achieve recovery is a feat that unsettles our assumptions about ancient medicine as crude or purely superstitious. It illustrates how observation—seeing that a survivor fared better than one untreated—could establish practices that endured across generations.

While trepanation highlights the extremes of early surgical courage, prosthetics reveal a quieter, yet equally profound, strand of medical innovation: the desire not only to preserve life but to restore function and dignity. Egyptian tombs have yielded artificial toes made of wood and leather, carefully shaped to fit a sandal. In some cases, wear patterns suggest that these were not ceremonial objects for burial but functional devices used in daily life. Similarly, a Roman bronze leg found in Capua, dating to around 300 BCE, demonstrates mechanical sophistication with joints designed to mimic natural movement. These artefacts remind us that medicine was not only about treating wounds but also about enabling people to return to social and economic roles. In a society where physical capacity often determined livelihood, prosthetics were more than technical curiosities; they were reintegrative tools. *The impulse to repair the body, rather than merely accept its losses, speaks to a fundamentally human refusal to surrender to limitation.*

Yet surgery without pain relief would have been unbearable, and here again we find evidence of pragmatic experimentation. Across cultures, healers discovered plant-based substances with anaesthetic or sedative properties. The Sumerians used opium poppies, the Chinese employed mixtures of hemp, and the Greeks relied on mandrake, henbane, and wine. These herbal preparations were not uniformly effective and carried risks of overdose, but they reflect a clear understanding of pharmacological action. The Egyptian medical papyri describe concoctions to dull pain during operations, while Indian Ayurvedic texts discuss the use of wine and cannabis for similar purposes. Accounts from Chinese sources even describe Hua Tuo, a physician of the Han dynasty, creating a compound anaesthetic known as *mafeisan*, reportedly enabling him to perform abdominal surgery. Though historians debate the details, the repeated appearance of such remedies across distant regions points to a shared drive to control suffering, guided by empirical trial and error.

These achievements invite us to consider what "experimentation" meant in pre-modern contexts. Without controlled trials or laboratories, knowledge advanced through observation, repetition, and adaptation. If a patient survived a skull surgery, that outcome reinforced the procedure's legitimacy. If an herbal mixture reduced agony, it became part of the healer's toolkit. This was science in its earliest, most tentative form: iterative, uncertain, yet cumulative. It is tempting to judge ancient surgery against today's sterile standards and find it wanting, but to do so misses the more important truth—that experimentation, even in rudimentary form, was at the heart of medicine from the beginning.

At the same time, the risks were immense. Infection could easily follow any incision, and there is ample evidence of procedures that failed catastrophically. Some trepanned skulls show no regrowth, signalling death during or immediately after the operation. Herbal sedatives could cross the narrow line between easing pain and causing fatal respiratory collapse. Even prosthetics, if poorly fitted, might have caused more suffering than relief. Ancient healers operated in a world of constant tension between necessity and danger. This tension did not deter them; rather, it seems to have spurred the refinement of techniques and the codification of practices in medical texts.

The Greek physician Hippocrates, often celebrated as the "father of medicine," discussed trepanation in his writings with striking detail, recommending it for skull fractures but advising caution to avoid injury to the brain. Such instructions show that surgical knowledge was not left to individual guesswork but transmitted, criticised, and improved across generations. Similarly, Indian surgical texts like the *Sushruta Samhita* contain elaborate descriptions of procedures ranging from cataract extraction to rhinoplasty, complete with specifications for instruments and postoperative care. Sushruta's emphasis on cleanliness, even without knowledge of microbes, suggests a proto-awareness of the link between hygiene and outcomes. His insistence that surgeons should practise on vegetables and animal parts before attempting operations on humans also points to a culture of systematic training, an early precursor to today's surgical residencies.

The existence of these surgical traditions also complicates the assumption that ancient medicine was entirely bound to ritual or religion. Ritual elements certainly framed many procedures, and spiritual explanations for illness abounded. Yet the practical details—choice of instruments, dosage of anaesthetics, methods of wound closure—reveal a pragmatic layer that cannot be dismissed as symbolic. Ritual may have provided meaning, but experimentation provided survival. *The two were not mutually exclusive; instead, they coexisted in a way that allowed medicine to serve both the body and the spirit.*

Looking at these practices from today's vantage point, it is easy to marvel at their boldness. But it is equally important to see their continuity with our own era. Modern surgery is still a delicate balance of risk and reward, carried out with tools that are more advanced but guided by the same principles: precision, training, pain management, and the hope of restoring function. The persistence of prosthetics into the twenty-first century, now enhanced by robotics and neural interfaces, echoes the same impulse seen in ancient Egypt—to restore what injury or disease has taken. The enduring use of plant-derived anaesthetics, from morphine to modern derivatives, continues a lineage that began when healers first noticed the numbing effects of opium or mandrake.

The history of surgery and experimentation reminds us that medicine is not a straight march of progress but a series of attempts, failures, and refinements. Ancient healers did not have microscopes or sterile theatres, but they had courage, curiosity, and an unrelenting drive to reduce suffering. Their legacy is not measured only in the bones that healed or the prosthetics worn but also in the mindset they embodied: that intervention, however risky, was preferable to resignation. In that sense, their experiments laid the groundwork not just for technical innovation but for the very ethos of medicine itself.

Knowledge Through Ritual

The earliest forms of medicine cannot be disentangled from ritual. For communities without microscopes, laboratories, or formal physiology, health and disease were framed within the symbolic order of their culture. Illness was rarely seen as a mechanical failure of the body alone; it was interpreted as imbalance, misalignment, or breach in the relationship between human beings, their community, their ancestors, and the cosmos. Rituals were not merely decorative ceremonies appended to treatment—they were treatment. They ordered experience, gave meaning to suffering, and mobilised the collective in acts of healing. Modern medicine, for all its empirical power, sometimes forgets how central meaning is to the experience of being unwell. The ancient reliance on ritual reminds us that knowledge of the body was always knowledge of the world, and both were approached together.

Anthropological evidence from societies across continents shows how healers drew on ritual to diagnose, to comfort, and to cure. Among the Navajo, the elaborate sand painting ceremonies were not only spiritual acts but diagnostic and therapeutic interventions. The patient's illness was reinterpreted as a disordered relationship to the cosmic order, and the ritual sought to restore harmony. Whether or not one accepts the cosmology, the effect was palpable: patients felt part of a larger story, their suffering given purpose, their recovery supported by the shared belief of the community. This is not a minor point. *The placebo effect, so often seen as a curiosity of modern trials, demonstrates in controlled conditions what ritualised healing embodied for millennia: that meaning and expectation are themselves potent medicines.*

Egyptian healing practices offer another example of ritual fused with knowledge. Temples dedicated to Imhotep or Asclepius functioned as both spiritual sanctuaries and centres of medical learning. Patients would sleep within the temple precinct, awaiting dreams sent by the gods that were then interpreted by priests as diagnostic tools. Herbs and surgical treatments were administered alongside invocations and symbolic acts. The ritual did not obscure the practical measures; rather, it anchored them in a framework that was culturally comprehensible and emotionally sustaining. Healing was not reduced to a technical procedure—it was an event that bound the individual to the divine and the community.

The role of ritual was not uniform, nor was it naïve repetition. It was a way of testing and transmitting knowledge in contexts where empirical experimentation alone was insufficient or too fragmented. In ancient China, the theory of yin and yang and the five elements was not just an abstract philosophy but a guiding system for diagnosis and treatment. A healer reading a patient's pulse did so within the logic of balance and correspondence, reinforced by rituals that linked human rhythms with cosmic cycles. These frameworks were not random inventions; they encoded generations of careful observation into symbols that could be taught, remembered, and ritually enacted. *Ritual served as both mnemonic and method, preserving complex medical insight in forms that were stable and transmissible before the age of print.*

Critics might argue that ritual-based medicine was a barrier to progress, binding practitioners to symbolic interpretations that obscured biological mechanisms. There is truth in this. The persistence of bloodletting in Europe, grounded in the ritual logic of balancing humours, undoubtedly caused harm and delayed the acceptance of circulation theory. Yet to stop at this critique is to miss the broader picture. Ritual provided a way of making sense of phenomena that had no clear explanation and often no cure. By placing illness within a cultural narrative, it mitigated despair and created a space where healing—whether physical, psychological, or social—could occur. Even ineffective rituals could have profound therapeutic effects by alleviating fear and fostering community solidarity.

There is also the question of authority. Rituals were not only about healing but about legitimising the role of the healer. Shamans, priests, or physicians were more than technicians; they were custodians of cosmological order. Their authority was enacted through ritual. To heal without ritual would have stripped their practice of its social recognition. Authority was itself therapeutic, reinforcing the patient's belief in the treatment and ensuring compliance. The chant, the symbolic gesture, the sacrificial offering—these were not peripheral; they were integral to the act of care, binding the healer, the patient, and the wider society into a shared process of recovery.

Modern medicine still retains vestiges of this dynamic, though often unacknowledged. The white coat, the stethoscope, the hospital gown—all carry symbolic weight, shaping how patients perceive authority and trust. The structured encounter of diagnosis, prescription, and follow-up is as much ritual as it is procedure. Clinical trials themselves, with their carefully designed protocols, blinding, and placebos, are modern rituals of verification. What the ancients enacted through prayer and symbolic acts, we now perform through statistics and protocols. Both serve to establish legitimacy, to provide order, and to persuade communities to invest belief in the outcome.

The coexistence of ritual and empirical practice was not accidental; it reflected an understanding that healing is always more than biology. In India's Ayurvedic tradition, treatment combined precise dietary regimens and herbal formulas with mantras and ceremonies designed to realign the patient's spiritual state. In sub-Saharan Africa, divination rituals identified not only the physical cause of illness but also the social or spiritual breach believed to underlie it, ensuring that the cure restored harmony at every level. These practices remind us that healing has always been both material and symbolic, practical and metaphysical.

What stands out across these cultures is the adaptability of ritual. Rituals were not static performances; they evolved, absorbed new knowledge, and adapted to empirical observations. A ritual that consistently failed might be abandoned or reinterpreted. Remedies that proved effective were folded into the symbolic framework, gaining legitimacy through ritual use. In this way, ritual was not opposed to experimentation but provided the social and cultural vessel through which experimentation could be integrated and preserved. *Without ritual, knowledge would have been too fragile to transmit; with it, knowledge could endure and be refined.*

The lesson for us today is not to romanticise ancient rituals but to recognise their role in shaping how societies understood and acted upon illness. They remind us that healing is never just the manipulation of tissues or chemicals; it is also the restoration of meaning, order, and trust. To view ancient ritual medicine as mere superstition is to underestimate the sophistication with which human beings have always sought to knit together body, mind, and culture in the pursuit of health. The power of ritual, whether in a Navajo sand painting, a Greek temple incubation, or a Chinese pulse reading, lay in its capacity to make suffering intelligible and recovery imaginable. That, in itself, is a form of knowledge worth acknowledging.

Trial, Error, and Adaptation

One of the most remarkable features of early medicine is not what it lacked but how it persevered. Without germ theory, molecular biology, or systematic anatomy for much of human history, healers were still compelled to act. People bled, bones broke, fevers raged, and childbirth claimed lives. Against these inevitabilities, trial and error became not just a method but a survival imperative. The history of medicine, when stripped of retrospective romanticism, is a history of countless experiments carried out in the most precarious of laboratories: the human body in distress. That so many practices endured long enough to form traditions is itself evidence of an underlying, if slow, process of empirical adaptation. It was not the elegance of theory but the stubbornness of iteration that allowed ancient medicine to advance.

Evidence of this practical empiricism is scattered across cultures. Consider the use of willow bark in ancient Greece and China. Its bitter extract was noted to reduce pain and fevers long before modern chemistry identified salicin, the forerunner of aspirin, as the active compound. The persistence of its use across centuries cannot be explained by coincidence or symbolic logic alone. It was observed, tested on repeated occasions, and retained because it worked. The method was unsystematic, but the results were recorded in cultural memory. The pattern here is instructive: ritual and philosophy framed the use of remedies, but *it was experience—whether the patient improved or worsened—that ultimately determined whether a practice survived.*

Trepanation, the act of drilling or scraping a hole into the skull, offers another striking example. It is easy for modern observers to dismiss it as barbaric, yet archaeological evidence shows patients often survived the procedure. The frequency of healed cranial bone suggests that practitioners learned, through trial, which techniques and instruments increased survival. While their rationale may have invoked spiritual explanations—releasing evil spirits or restoring balance—the adjustment of technique reflects an empirical core. Those who drilled too deeply killed their patients.

Those who refined their method preserved life, and that knowledge was passed on. *Even when cloaked in ritual, practical outcomes shaped medical continuity.*

The same pattern is visible in the domain of childbirth. Across the ancient Mediterranean and Near East, midwives developed techniques to turn breech babies, deliver the placenta, and manage haemorrhage without any formal anatomy. Their knowledge was acquired not from dissection or theory but from generations of watching, attempting, failing, and adjusting. Herbal remedies for easing labour pain or encouraging contractions were retained because midwives observed patterns of efficacy. In cultures where written transmission was scarce, oral tradition carried these cumulative observations forward. What appears to us as folk wisdom was, in truth, an evolving record of collective experiment.

The absence of scientific theory did not prevent forms of testing that resemble clinical logic. In Ayurveda, remedies were often evaluated by observing their effects on doshic balance, but those that consistently harmed rather than healed tended to fall away. In Mesoamerican medicine, healers working with plants like peyote or cacao did not always understand the pharmacological basis but experimented with dosages, combinations, and contexts of use. Failures were remembered as cautionary tales, successes ritualised into treatment practices. Here the tension between symbolic interpretation and practical observation is evident: rituals preserved meaning, but *the patient's condition was the ultimate arbiter of credibility.*

The story of herbal anaesthetics illustrates this interplay particularly well. Mixtures containing opium, mandrake, or henbane appear in Greek, Roman, and Middle Eastern texts. Their application in surgery was hardly uniform—dosage varied wildly, risks of poisoning were high—but over centuries practitioners refined both preparation and administration. Galen's descriptions of "soporific sponges," cloths soaked in herbal mixtures and inhaled by patients, testify to this adaptive process. The practice was imperfect, but it reveals the careful adjustments of a craftsperson learning by consequence. Theories about humours or spirits may have framed the explanation, yet the incremental refinement of technique was unmistakably empirical.

We can see this more starkly by contrasting practices that endured with those that faded. Bloodletting persisted for centuries, but its overuse was often contested within its own tradition. Some physicians advocated moderation, others abandoned it when outcomes proved consistently poor. Leech therapy, however, persisted in part because patients occasionally showed visible improvement in swelling or infection. The persistence of one and the decline of another demonstrate a process of internal evaluation, however crude. Medicine was not static; it was contested terrain where experience shaped survival.

Importantly, this process was rarely linear. A discovery in one culture might remain localised for centuries until rediscovered elsewhere. The use of digitalis from foxglove in Europe for heart conditions appeared sporadically in folk practice long before William Withering systematically studied it in the eighteenth century. The fact that foxglove remained in use suggests that its efficacy was too evident to ignore, even if its dangers limited adoption. Ancient medicine advanced not through a continuous trajectory of progress but through waves of adaptation, in which small, hard-won gains were folded into broader traditions.

Philosophically, the reliance on trial and error speaks to a fundamental tension in human knowledge: the need for action in the absence of understanding. Theoretical clarity is a luxury; suffering demands intervention. Healers could not afford to wait for germ theory to sterilise wounds or for molecular biology to classify compounds. They had to act with what they had, and what they had was observation. If a root soothed pain in one patient, it might in another. If a procedure saved one life, it was attempted again. The process was crude, but it was not irrational. It was a pragmatic empiricism, shaped by necessity.

Modern parallels help us appreciate this continuity. Many frontline medical innovations still emerge from practice before theory. The discovery of lithium's effect on bipolar disorder preceded any understanding of its mechanism. The use of penicillin spread before its molecular action was fully elucidated. Emergency medicine often adopts techniques long before controlled trials confirm them, guided by repeated observation of outcomes. In this sense, the ancient reliance on trial and error is not as alien as it first appears. *The pattern endures: practice often outruns theory, and theory catches up later.*

The danger, then as now, lay not in trial and error itself but in dogma that refused to let outcomes reshape belief. Where healers remained open to adaptation, medicine advanced. Where symbolic or religious frameworks overruled experience, harmful practices persisted. The lesson from antiquity is not that trial and error was primitive but that adaptability was the hinge of progress. Knowledge that survived was knowledge that adjusted, even when explanations lagged behind. In that respect, ancient medicine was less a catalogue of superstitions than a living laboratory, constantly recalibrated by the stubborn fact of the human body's response.

If the modern reader looks back with impatience at the slowness of ancient medical progress, it is worth remembering that our own advances rest on the same foundation: a willingness to act in uncertainty, to learn from failure, and to adapt until the patterns coalesce into something more enduring. Ancient healers, without molecules or microscopes, modelled this principle with a persistence that ensured medicine survived long enough for science to emerge. Their legacy is not just the remedies that worked but the method that underpinned them: a relentless, if fragile, commitment to adjust when reality contradicted belief. That is the true continuity of medicine across the ages.

Chapter 8
Energy Without Engines

Long before coal furnaces roared and turbines spun, human societies discovered ways to channel the forces flowing through their environment. These were not engines as we understand them today, but rather adroit harnessings of fire, water, wind, and muscle that transformed daily life. To trace these early engagements with energy is to appreciate that industrialisation did not arrive fully formed in the eighteenth century; it was preceded by millennia of experimentation, a slow apprenticeship in persuading nature to cooperate. What is striking is not merely the technical ingenuity but the intellectual curiosity behind it: early thinkers did not simply exploit power, they wrestled with the very nature of motion, pondering whether work could persist indefinitely and how forces might be directed with consistency and purpose.

The first and perhaps most profound step was fire. To conceive flames as a controllable form of energy rather than an unpredictable hazard required a subtle cognitive shift. Once fire was confined to hearths and transported in embers, it became a transformative tool. Cooking unlocked nutrition otherwise unavailable; smelting extracted metals from stone; pottery kilns hardened clay, extending storage and trade. Even more subtly, fire extended the day: with illumination came evenings of labour, discussion, and communal gathering. This was not merely technical mastery but a reframing of perception: fire could serve humanity, not merely threaten it. In this reframing lies the origin of deliberate energy management, the first intellectual leap towards harnessing natural forces.

Water and wind then extended this mastery into mechanical motion. The waterwheel, deceptively simple in appearance, represented a profound abstraction. A river is capricious, its flow varying with rain, season, and geography. To divert and channel it into consistent rotation required seeing chaos as something that could be structured. Early wheels ground grain, lifting drudgery from human and animal muscle. Soon, they powered pumps, saws, and presses. The significance lay not only in what the wheel accomplished but in the conceptual leap: the translation of fluid, irregular motion into a rhythmic, predictable sequence.

Windmills, flourishing where rivers proved unreliable, achieved the same end by different means. From Persia to northern Europe, they drove irrigation, drained marshlands, and milled grain. Each design was custom-built to local wind conditions, an elegant acknowledgment that power could be borrowed from the unseen atmosphere. These machines did more than perform work; they demonstrated an emerging comfort with abstracting natural forces into mechanical sequences, a philosophy that would quietly underpin industrialisation centuries later.

The domestication of animals, too, was an act of energy engineering. Horses, oxen, and other draft animals were not mere beasts of burden but integral nodes in an emerging energy economy. Fields had to supply both humans and animals; stables, veterinary knowledge, and harnessing technologies developed in tandem. The plough pulled by oxen changed not only agricultural scale but temporal rhythm, creating surpluses that freed labour for other pursuits. Cavalry reshaped warfare, and pack animals extended trade across deserts and mountains. Societies treated these animals as predictable, maintainable engines within a larger infrastructure—a biological energy network requiring planning, breeding, and resource allocation.

Underpinning all these innovations was the intellectual fascination with motion itself. Once humans realised that fire, water, wind, and animals could be made to do work, they asked: what sustains motion, and could it ever be perpetual? Greek philosophers, from Aristotle onwards, speculated about natural causes and the persistence of movement. Medieval thinkers dreamed of wheels that might turn endlessly without fuel. These failed contraptions were not idle curiosities but earnest attempts to grasp energy's essence. The obsession with perpetual motion was as much metaphysical as mechanical: could a force be captured that required no replenishment? Though invariably frustrated by friction and resistance, these pursuits cultivated insights into energy conservation long before the principle was formalised.

Taken together, these threads show a progression from elemental control to systematic harnessing. Fire taught storage and direction; water and wind revealed the conversion of irregular flows into repeatable cycles; animals demonstrated the integration of biology into energy economies; philosophical inquiry probed the principles underlying motion itself. Steam engines did not emerge in isolation—they were anticipated by centuries of practice, reflection, and experimentation that trained societies to view energy as a resource to be disciplined, measured, and deployed.

The relevance today is unmistakable. Modern debates about renewables, sustainability, and the limits of growth echo the dilemmas of our predecessors. Just as medieval engineers strained against perpetual motion, we must recognise that no energy comes without cost. Just as early farmers balanced land for oxen and humans alike, we negotiate between crops, biofuels, and habitation. Today's turbines and hydroelectric dams are heirs not merely in form but in mindset to waterwheels and windmills: instruments through which humanity learns to borrow, channel, and regulate natural forces.

This chapter, therefore, is not a catalogue of quaint devices but an inquiry into the origins of energy literacy. Each story—of fire, water, wind, animals, and the intellectual reflection they inspired—reveals a quieter, slower, yet no less profound revolution. Industrialisation did not appear ex nihilo; it emerged from a long apprenticeship in thinking about energy as a resource, imagining machines as intermediaries, and reconciling human ambition with nature's laws. The lesson is enduring: to control motion is to shape civilisation, and that lesson began long before the first engine turned.

Fire as Controlled Power

Few innovations in human history rival the mastery of fire for its transformative impact. The ability to kindle, sustain, and direct flames was not simply a technological breakthrough but an intellectual one: it redefined humanity's relationship with nature, converting a force once feared as destructive chaos into a controllable source of nourishment, protection, and power. Fire was not discovered in the sense of a tool ready-made in the environment, like a stone or a branch; it had to be reimagined. To live alongside fire without dread, to carry embers, to kindle sparks, and eventually to use it deliberately for tasks both ordinary and monumental required an audacious reframing of risk into opportunity. *That mental shift, from danger to tool, was one of the earliest and most profound examples of humans seeing the world not just as it was, but as it could be shaped.*

Archaeological traces of controlled fire reach back hundreds of thousands of years. Evidence from sites such as Wonderwerk Cave in South Africa, dated to roughly a million years ago, suggests that early hominins were not only exposed to fire but learned to maintain and exploit it. Charred bones and ash deposits show the intentional use of fire for cooking, not merely the incidental remains of a wildfire. Cooking was more than a convenience; it altered the human body itself. Heat transformed fibrous roots, grains, and meat into digestible forms, releasing calories otherwise locked away. This had biological consequences: shorter digestive tracts, smaller teeth, and larger brains, all made possible by a higher yield of accessible energy. In this sense, fire was not just a tool of survival but an engine of evolution.

Yet controlling fire was never trivial. Wildfires could not be summoned on demand, and lightning was unreliable. Early communities had to cultivate skill in tending embers, transporting them, and nurturing them into flame. Some groups learned to generate fire through friction or flint, but more often it was preserved, not produced. To guard fire was to guard life itself. It shaped patterns of settlement, social roles, and rituals of vigilance. The hearth became not merely a physical centre of camp or dwelling but a symbolic one, around which stories were told, tasks shared, and societies anchored. Fire transformed evenings into communal space, extending the day beyond sunset and enabling the creation of cultural bonds as much as physical sustenance.

The utility of fire quickly expanded beyond the hearth. One of its earliest great uses was in shaping landscapes. Indigenous peoples across continents practised controlled burning to clear underbrush, encourage the growth of desired plants, and manage animal populations. These techniques demonstrated that fire could be not only endured or harnessed in contained spaces but applied on a large scale to manipulate ecosystems. Far from being primitive, this was a sophisticated ecological management strategy, one that anticipated modern forestry and agriculture. *Fire, wielded with precision, was an instrument of planning rather than accident.*

In metallurgy, fire reached a new threshold of significance. Smelting ore into copper, bronze, and eventually iron marked the shift from Stone Age to metal age. This required not only high heat but sustained and concentrated heat, achieved through innovations such as furnaces and bellows. The very idea that stone could be transformed into liquid and then shaped into tools was a conceptual leap built on trust in fire's power to transmute matter. The blacksmith's forge became an emblem of civilisation, where raw nature was reshaped into artefacts of war, agriculture, and artistry. To master metallurgy was to believe in fire's potential as a creative force, not only a destructive one.

Ceramics tell a similar story. Clay, mouldable but fragile, was remade by fire into pottery durable enough to store grain, water, and oils. Kilns, with their carefully regulated temperatures, show a refined understanding of how fire could be modulated and directed. It was not brute burning but the nuanced management of heat that distinguished skilled potters. The leap here was not only technological but conceptual: fire could be graduated, measured, and tuned to yield entirely different outcomes. A poorly controlled flame would crack pots; a well-managed kiln produced vessels that supported trade, cooking, and long-term storage. The subtlety of this knowledge reflected generations of empirical learning, a form of science before theory, with fire at its centre.

Religious and philosophical traditions also testify to the deep intellectual weight fire carried. In Zoroastrianism, fire was revered as a sacred element embodying purity and divine order. In ancient Greece, the myth of Prometheus, who stole fire from the gods, signalled recognition of its transformative capacity as both gift and transgression. In Vedic rituals of India, fire mediated between humans and deities, consuming offerings and carrying them upward. These traditions capture an essential truth: fire was never a neutral tool. It evoked awe, demanded respect, and symbolised both creation and destruction. *To control fire was to hold a fragment of divine power, and this perception both restrained and inspired its use.*

If fire's earliest contributions were in survival, nourishment, and culture, its later applications foreshadowed industrialisation. Charcoal, produced by slow-burning wood, became a critical fuel for smelting iron on larger scales. Fire drove early engines of chemistry, from distillation to the making of glass. The intellectual transition from fire as a living force to fire as a form of energy was slow, but it was decisive. Once seen as a spirit or element, fire was increasingly understood as a process—combustion—that could be harnessed systematically. This reconceptualisation paved the way for the later sciences of thermodynamics and mechanics.

Modern analogies reveal just how radical this transformation was. Imagine electricity, with its dangers of shocks and fires, being first encountered without any infrastructure of safety or explanation. To turn such a hazard into a ubiquitous utility required a complete reframing of perception, infrastructure, and practice. Fire underwent precisely this transformation. It became a managed, predictable, almost invisible partner in daily life, so domesticated that modern societies often forget its wild origins. Yet every stove, furnace, and matchbox still carries within it that ancient negotiation: the taming of volatility into reliability.

The intellectual leap was therefore not only in how fire was used but in how it was thought about. Early humans began to distinguish between raw nature and manageable force. They developed categories of *dangerous* versus *useful*, destructive versus sustaining. This capacity to reclassify phenomena, to reimagine them as potential rather than threat, was a mental habit that underpinned countless later discoveries. Fire taught humanity that the environment was not fixed but could be reshaped through understanding and persistence. It taught that nature's most feared forces could be enlisted as partners in human endeavour.

Today, when discussions of energy focus on electricity grids, nuclear power, and renewables, it is easy to overlook that the foundational lesson of energy management was first learned with fire. Its controlled use forced humanity to grapple with fundamental questions: how to balance risk and reward, how to organise labour around a shared resource, how to preserve continuity of supply, and how to imagine the future as more than mere survival. *The leap from fearing flames to wielding them was not just technological but cognitive; it was one of the earliest demonstrations of humanity's unique capacity to see power in peril and to make knowledge from danger.*

Waterwheels and Windmills

When societies first realised that natural forces could be captured and converted into usable motion, they stumbled upon one of the most enduring foundations of technological progress. The waterwheel and the windmill were not spectacular inventions in the sense of sudden, revolutionary discovery; rather, they represented the slow crystallisation of a profound idea—that energy did not have to come from muscle, whether human or animal, but could be drawn continuously from the environment. This conceptual leap created a lineage that stretches directly to industrialisation, modern turbines, and the global energy systems of today. *The story of waterwheels and windmills is therefore less about quaint rural machinery than about humanity's first serious attempt to mechanise power on a scalable basis.*

The waterwheel appears in records as early as the Hellenistic period. Ancient Greek engineers, such as Philo of Byzantium and Vitruvius, described vertical wheels turned by flowing water to grind grain. Rome expanded this principle, building mills along aqueducts and riverbanks to supply flour to urban populations. The Barbegal mill complex in southern France, dated to the second century CE, is perhaps the most striking example: sixteen waterwheels arranged in series, capable of producing enough flour to feed a large city. This was industrial production long before the industrial revolution. The scale and organisation reveal not only mechanical ingenuity but also a new vision of labour—shifting the burden from countless backs and hands to a persistent flow of water harnessed by wood and stone.

Wind power developed somewhat later, first attested in Persia around the seventh century CE, where vertical-axis windmills ground grain and pumped water. The design spread westwards, with Europe adopting horizontal-axis windmills by the twelfth century. These became iconic features of the medieval landscape, dotting the Netherlands, England, and Spain. But behind their picturesque silhouettes lay a serious economic role. They drained wetlands to reclaim farmland, powered sawmills, and ground grain on scales that transformed local economies. In each case, wind was turned from a force that battered ships and eroded landscapes into a resource that sustained communities. *The act of turning wind into work was nothing less than a redefinition of what counted as a reliable source of power.*

What made both waterwheels and windmills intellectually significant was not only their mechanics but the broader principle they embodied: the conversion of continuous natural motion into repetitive, useful mechanical tasks. Unlike the fire-based technologies of metallurgy or pottery, these machines offered not heat but motion, and motion could be applied to a multitude of tasks. Grain grinding was the most common, but the same rotary energy was adapted to hammers, bellows, saws, and textile mills. Each application represented an act of abstraction: energy became interchangeable. One did not need to think of "a waterwheel for grain" or "a windmill for wood"; one thought of captured motion as a general-purpose power, waiting to be channelled into different machines.

Philosophically, this was a decisive shift. Before such devices, labour was measured primarily in human or animal terms: how many people could turn a quern, how many oxen could plough a field. Water and wind broke that equivalence, suggesting that power was not limited by biology. Instead, it could be scaled by engineering. A stronger current meant a faster wheel; a taller windmill with broader sails caught more air. The limiting factor became not muscle but design. *This relocation of power from flesh to environment marked the beginning of an intellectual divorce between labour and energy, a separation that industrialisation would later exploit fully.*

Yet adoption was neither uniform nor inevitable. In the Roman Empire, despite feats such as Barbegal, watermills never replaced slave labour on a large scale. Social historians argue that the abundance of enslaved workers reduced incentives to mechanise. By contrast, in medieval Europe, where labour shortages followed the Black Death, mills became essential. They supplemented a diminished workforce and gave landowners leverage over production. Thus, the spread of mechanical power was entangled not only with engineering but with social and economic structures. Technology was never neutral; it advanced where conditions demanded or rewarded it.

Culturally, mills also became sites of negotiation between communities and technology. In medieval England, "lord's mills" were often monopolies: peasants were required to grind grain there and pay a fee. The creak of the millwheel was therefore both the sound of sustenance and the sound of obligation. Resistance to such monopolies sometimes flared into conflict, illustrating that mechanical power, once centralised, could reshape social hierarchies as much as it relieved labour. *The waterwheel and windmill were not just technical devices but instruments of governance and control, revealing how technology and power—literal and political—were intertwined.*

The intellectual legacy of these early machines is visible in the very language of mechanics. Gears, cogs, axles, shafts—these were not abstractions from theoretical physics but practical solutions developed to transmit the irregular force of wind or water into steady motion. They laid the groundwork for mechanical engineering as a discipline. When Renaissance thinkers like Leonardo da Vinci sketched designs for elaborate gear trains and perpetual motion machines, they were building upon the mechanical vocabulary forged in mills. The fascination with motion, with its transfer and transformation, began with these humble devices.

There is a striking continuity between these pre-industrial machines and modern energy systems. A hydroelectric dam is, at its core, a waterwheel on a colossal scale, converting falling water into rotary motion to turn turbines. A wind farm is an array of windmills, their blades optimised by aerodynamics but still translating gusts into rotation. The principles remain the same; only the scale and materials differ. This continuity underscores the enduring insight: *to harness the environment is to amplify human capability beyond the constraints of muscle.*

Yet there is also a contrast worth noting. Pre-industrial mills were embedded in local ecologies. A village mill depended on the seasonal flow of its river or the regularity of coastal winds. Its power was sustainable, if limited. Industrialisation, in contrast, sought universality and predictability, favouring fossil fuels precisely because they could be stored, transported, and burned on demand. In this light, the waterwheel and windmill represent not only the precursors of industry but also an alternative lineage—a demonstration that renewable power can sustain economies when harnessed intelligently. The renewed interest in wind and water energy today, from tidal turbines to offshore wind farms, is not innovation ex nihilo but a return to an old, nearly forgotten wisdom.

There is a temptation to romanticise waterwheels and windmills as relics of a slower, gentler age. Yet to the societies that built and depended on them, they were anything but quaint. They were high technology, requiring careful siting, investment, maintenance, and expertise. They restructured economies, reshaped landscapes, and demanded new skills. They symbolised a willingness to trust machines with critical tasks once performed by human hands. Above all, they represented a growing confidence that natural forces, once capricious and intimidating, could be made into partners in production.

The lesson they leave us is both historical and philosophical. The leap from watercourse to wheel, from wind to sail, was not an isolated technical trick but an intellectual commitment to the idea that motion itself was a resource. *Once humanity saw energy not as a fixed inheritance of muscle but as a harvestable flow in the world around us, the pathway to industrialisation opened.* The waterwheel and windmill were thus not merely mechanical devices but milestones in the human reimagining of energy, standing at the threshold where nature's rhythms became civilisation's engines.

Animal Power as System

To describe domesticated animals merely as companions or as sources of food is to miss their deeper historical role: they were once central components of human energy economies. Before the mechanisation of industry and the harnessing of fossil fuels, oxen, horses, camels, donkeys, water buffalo, and even dogs were not just beasts of burden but essential converters of biological energy into work. *Their integration into systems of labour represented one of the most profound expansions of human capability prior to industrialisation, transforming agriculture, transport, warfare, and commerce.*

The domestication of animals was not in itself enough to create an energy system. A sheep or goat might provide wool and milk without altering patterns of power, but the harnessing of strength for traction and transport required a conceptual leap: to view animal muscle as a form of renewable energy, one that could be managed, scaled, and sustained through breeding and care. By training oxen to pull ploughs, humans extended the effective force available to cultivate land far beyond what a farmer's body alone could achieve. Yields expanded, fields deepened, and surplus production became possible on a scale that enabled cities and states to flourish. Here, the animal was not an individual helper but an integral part of a system in which grain, fodder, manure, and labour circulated in a closed loop.

Horses carried this system further, particularly once the invention of the horse collar in medieval Europe unlocked their full strength without choking them. Unlike oxen, horses could combine speed with power, enabling not only deeper ploughing but faster transport. This seemingly simple harness innovation transformed economies. Agricultural productivity increased, trade accelerated, and even postal systems became viable across large territories. The mobility afforded by horse power expanded political geography: empires could be governed across vast distances, military campaigns could be mounted with logistical sophistication, and cultural exchange quickened. *Technology in this context was not a machine of metal and gears, but a reconfiguration of biology guided by human insight.*

Camels offer a complementary example, tailored to their ecological niche. In arid environments where wheels bogged down and water was scarce, camels converted sparse vegetation into sustained transport across deserts. Their endurance made trans-Saharan and Silk Road trade networks possible, linking economies across continents. Unlike watermills or windmills, whose operation was tied to specific landscapes, animal power was portable: it could move with armies, caravans, or settlers, extending human reach wherever pastures could sustain it.

These animal systems also carried political and social implications. Control of draught animals could determine wealth and status. In agrarian societies, owning teams of oxen or horses conferred the ability to farm more land and produce more surplus, entrenching inequalities. Laws often reflected this centrality; theft of working animals could carry punishments as severe as theft of gold. In many cultures, tribute and taxation were collected not only in grain or coin but in livestock, recognising their role as units of stored energy. What counted as wealth was inseparable from what counted as power, in the literal sense of force.

The integration of animals into labour systems required infrastructure and organisation. Roads, pastures, watering points, and stables became part of the energy network. In medieval towns, for instance, the rhythm of commerce depended on the steady flow of cart horses bringing goods to market, just as much as on merchants or moneylenders. In China, water buffalo were so embedded in rice cultivation that rural calendars were structured around their cycles of labour and rest. *To study these societies is to see animals not as external aids but as co-workers, embedded in the architecture of daily life and the logic of economies.*

At the same time, reliance on animal power exposed human societies to vulnerabilities. Disease could devastate herds, crippling agriculture or transport overnight. The rinderpest epidemics of the nineteenth century wiped out millions of cattle across Africa, collapsing food systems and reshaping political orders. Dependence on living power was therefore never without risk, and managing that risk was as much part of the system as breeding or harness design. This reality underscores a point often forgotten in retrospect: animal-based energy was powerful but precarious, efficient but fragile.

Philosophically, the harnessing of animals raises enduring questions about the boundary between nature and technology. Were oxen and horses tools, in the same way as ploughs and carts, or were they partners in production? Ancient and medieval thinkers wrestled with this ambiguity. Aristotle classified animals as instruments of labour, yet many cultures granted them ritual or legal protections, acknowledging their unique status as living engines. The paradox lies in the fact that animals possessed will, fatigue, and needs of their own. Unlike water or wind, they resisted, learned, and sometimes refused. The energy system was therefore not entirely mechanical but relational, requiring training, care, and even forms of persuasion.

In a sense, animal power was the first hybrid technology: part biological, part cultural, part mechanical. Yokes, collars, saddles, carts, and reins were devices that mediated between human intention and animal muscle. The system functioned only through this interface. To imagine an oxen-driven plough without the ploughshare and yoke is as incomplete as imagining a windmill without sails. The intellectual achievement was not merely in domesticating animals but in designing the apparatus that linked their energy to human purposes.

Looking forward, one might be tempted to dismiss animal power as an obsolete stage, overtaken by engines and electricity. Yet its influence persists, both practically and conceptually. In many regions of the world today, donkeys, camels, and oxen still provide transport and traction. Even in highly mechanised societies, equestrian sports, working dogs, and cultural reverence for draught animals testify to the deep imprint of this history. More abstractly, the very concept of horsepower—still used to measure mechanical output—testifies to the enduring mental model of animals as the baseline unit of power. *Modern energy discourse continues to be shaped by this ancient calibration between muscle and machine.*

The transition away from animal-based energy systems was neither smooth nor uniform. In parts of Europe and North

America, mechanisation displaced horses and oxen within a century. In others, animal power remained essential well into the twentieth century. The story is not one of simple progress but of overlapping systems, where animals, water, wind, and fossil fuels all coexisted in varying proportions. Each system shaped not only economies but also ways of thinking about power, labour, and control over the environment.

Ultimately, the integration of animals into energy economies reveals a stage in human history where biology and technology merged most intimately. It was a time when progress was measured not in kilowatts or coal tonnage but in the stamina of oxen and the speed of horses. Yet the intellectual principle—*that power can be abstracted, stored, and applied systematically*—was the same principle that later underpinned steam engines and electricity. By treating animals as components in an energy network, humans began to think in systems rather than tasks, in flows of energy rather than isolated acts of labour. That conceptual shift was as significant as any technical invention, for it set the stage for the mechanised and industrialised world that followed.

Philosophy of Motion

The human fascination with motion is as old as recorded thought, and it carries with it one of the most enduring intellectual puzzles: can motion sustain itself indefinitely, or must it always be fed by an external source of energy? For ancient thinkers, this was not an idle curiosity but a pressing philosophical question bound up with the nature of the cosmos, the mechanics of the body, and the potential limits of technology. To wrestle with the idea of perpetual motion was to wrestle with the very logic of change, continuity, and causation. *The dream of a machine that could turn forever without fuel was more than a technical fantasy—it was a window into humanity's shifting understanding of nature and power.*

Aristotle set the early framework, insisting that motion required a cause: an object in motion would stop unless something kept it going. This view, rooted in everyday observation, seemed self-evident—after all, push a cart and it eventually halts. Yet implicit in his theory was a limitation that troubled later thinkers. If everything required a mover, then what sustained the heavens? Aristotle's solution was metaphysical rather than mechanical: he posited the existence of a "Prime Mover," a perfect and eternal cause of celestial motion. In this early form, perpetual motion was considered real, but only in the heavens, where divine order governed. On Earth, friction, resistance, and decay ruled out the possibility of endless self-sustaining motion. This cosmological division between the earthly and the celestial lingered for centuries, shaping the boundaries of what was thought possible.

By the Middle Ages, the notion of perpetual motion had taken on a more mechanical character. Philosophers and inventors across Europe, the Middle East, and Asia devised ingenious contraptions: wheels laden with asymmetrically weighted arms, fountains designed to recycle their own water, and magnetic devices that promised unceasing rotation. None of these worked as intended, but their proliferation tells us something significant. *The idea of perpetual motion persisted not because of naivety, but because it exposed a gap between observed phenomena and the conceptual tools available to explain them.* If the stars moved endlessly, why not a wheel on Earth? If rivers seemed never to tire, why could their motion not be captured in an infinite loop? The fascination with perpetual motion was less about building practical machines than about interrogating the limits of natural law.

The turning point came with the gradual emergence of a more precise physics of motion. Islamic scholars such as Avicenna introduced the concept of *impetus*—the idea that an object, once set in motion, carries an internal force that sustains it until something dissipates it. This was a clear departure from Aristotle and foreshadowed Newton's formulation of inertia centuries later. In this light, the perpetual motion debate shifted: motion itself could persist without constant pushing, but energy could not be endlessly recycled without loss. Friction, drag, and inefficiencies acted as thieves of motion, dissipating energy into heat. The impossibility of perpetual motion began to emerge not as a matter of divine decree but as a consequence of measurable physical laws.

By the seventeenth century, the rise of experimental science gave new clarity. Galileo, through his studies of inclined planes, realised that in the absence of resistance, a ball would roll indefinitely. Newton codified this in his First Law of Motion: a body in motion remains in motion unless acted upon by an external force. This was, in a sense, a partial vindication of the dream of perpetual motion. Motion itself was not self-extinguishing. What doomed perpetual motion machines was not motion's fragility but energy's dissipation. The concept of conservation of energy, sharpened in the nineteenth century, formalised this limit. A machine cannot create more energy than it consumes, and some energy is always lost to heat, sound, or wear. Thus perpetual motion was not merely unachievable in practice but impossible in principle.

Yet this impossibility did not dampen curiosity—it redirected it. Philosophers of motion began to focus not on eternal machines but on the transfer and transformation of energy. Windmills converted air currents into mechanical rotation; waterwheels transformed falling water into grinding force; animals turned grass into traction. Each system displayed the same principle: energy could be moved, repurposed, and stored, but never conjured from nothing. The fascination with perpetual motion thus yielded to a more profound realisation: *the world was governed not by fanciful devices but by hidden balances, trade-offs, and limits.* This insight laid the groundwork for thermodynamics, one of the cornerstones of modern science.

The cultural persistence of perpetual motion schemes, however, should not be dismissed as folly. They reveal a psychological truth: humans long to transcend scarcity. An engine that never stops is a metaphor for freedom from toil, dependence, and decay. The countless attempts to design such devices—whether through overbalanced wheels, siphoning fountains, or self-turning screws—speak to a deep desire for autonomy over nature's constraints. Even as physicists proved the impossibility, charlatans continued to exploit the fascination, and dreamers continued to sketch new designs. This tension between fantasy and law is part of what makes the history of motion so compelling: it is as much about human hope as it is about physical limits.

Modern echoes of this debate can be found in how we speak about renewable energy. Solar panels and wind turbines are not perpetual motion machines, but they are sometimes framed as if they were—a promise of endless power from inexhaustible sources. The philosophical caution is the same: energy may be abundant, but it must still be captured, converted, and transmitted within the bounds of thermodynamics. Perpetual motion may be impossible, but the drive to approximate it—by reducing loss, extending efficiency, and drawing from flows that seem eternal—continues to shape technological ambition.

One can also read the philosophy of motion as a lesson in intellectual humility. For centuries, great minds pursued an impossible goal, only to discover that the impossibility itself was illuminating. By confronting what could not exist, thinkers refined their understanding of what could. The impossibility of perpetual motion led directly to the conservation laws that underpin modern physics, just as the dream of flight spurred the aerodynamics that eventually made it real. *Limits, once grasped, often generate the deepest insights.*

What unites the long history of thought about motion is a gradual but profound shift: from viewing the world as a set of isolated phenomena to conceiving it as a system of energy transfers. To the ancient farmer, a cart stopped because the ox grew tired; to Newton, it stopped because friction stole its motion; to modern physics, it stopped because energy diffused into less usable forms. Each stage reflects not only a refinement of theory but a widening of imagination. The philosophy of motion is thus less about unattainable machines and more about the intellectual courage to pursue a problem to its very limits—and to accept that in those limits lies the key to deeper understanding.

Chapter 9
Cities as Collective Inventions

Cities are among humanity's boldest experiments. They are far more than clusters of buildings or dense populations; they are deliberately orchestrated environments, designed to channel power, encode values, and shape the rhythms of collective life. The emergence of the city marks a threshold: people were no longer organising solely around kinship, tribe, or immediate ecology, but within artificial frameworks that demanded negotiation, compromise, and design. What sets cities apart from towns or villages is not scale alone but intention—a conscious attempt to construct a shared system of living that endures beyond immediate survival, a physical manifestation of how humans imagine communal life. To study ancient cities is to see philosophy and politics rendered into stone and space, producing infrastructures that were practical, yes, but also ideological.

The grid is one of the most striking expressions of this ambition. From Mohenjo-Daro in the Indus Valley to the disciplined streets of Roman colonies, street layouts reveal a logic of order. A grid is a deliberate choice, not a necessity. It signals a belief that space can be rationalised, subdivided, and standardised, that life itself can be arranged according to principle rather than left to organic sprawl. Grids suggest predictability and equality: dwellings front similar streets, blocks follow the same logic. They are convenient for drainage and movement, certainly, but their significance runs deeper. Grids inscribe authority into the very ground, reflecting rulers' desire to codify control. In Chinese imperial capitals, avenues aligned with the cardinal directions, linking earthly governance to celestial order.

When one walks Manhattan's streets or Barcelona's Eixample, one is not merely navigating space but inhabiting a philosophy: that repetition and order produce stability, fairness, and clarity.

If streets encode political logic, infrastructure often embodies ethical and philosophical commitments. Aqueducts, for instance, delivered water over kilometres, a feat of engineering that required communal prioritisation. Roman aqueducts were not just practical but declarative: the city's lifeblood belonged to all, worth immense effort and expense. Sewers, too, were investments in civic health, recognition that what flows unseen beneath the streets determines life above them. Roads, while conduits of movement, also expressed connectivity, bringing provinces into the sphere of central authority, enabling trade, military logistics, and cultural exchange. Infrastructure is never neutral; choices about aqueducts versus walls, markets versus fortresses, reveal societal priorities and ideals about endurance.

Markets offer perhaps the clearest illustration of cities as systems of knowledge, not merely commerce. At first glance, they were spaces of buying and selling. Yet the agora in Athens, the Forum in Rome, and the bazaars of Damascus also hosted the circulation of ideas. Merchants carried news; artisans displayed techniques; citizens debated politics amid food and textiles. Information flowed through markets embedded in interaction, not abstract media. Dense proximity enabled a collective intelligence impossible in scattered villages. Innovations spread, reputations formed, decisions were tested in public view. In a sense, these markets functioned as the proto-internet of antiquity, physical architectures transforming local exchanges into networks of shared understanding.

Yet ingenuity carries fragility. Many early urban experiments collapsed spectacularly. The ruins of Angkor, Mohenjo-Daro, and Teotihuacan testify to the precariousness of urban systems. Water infrastructure failed under climatic stress; trade routes shifted or disappeared; densities that once fostered innovation also amplified vulnerability to disease, famine, and unrest. Urban construction tested the limits of collective organisation, and those limits often proved narrower than hoped. Collapse was rarely mere failure—it was often adaptation, a reminder that cities are living systems rather than static monuments. Their vulnerabilities underscore a paradox: the very concentration, resource centralisation, and value encoding that empower cities also expose them when conditions change.

The history of ancient cities, therefore, is not a simple narrative of triumph or decline but of experiment. Each urban form embodies a hypothesis: grids show the power of order, aqueducts the promise of shared resources, markets the intelligence of proximity, collapse the cost of overreach. Cities are not by-products of population growth but deliberate choices to structure interdependence. Through their streets, aqueducts, and marketplaces, ancient peoples left legacies that persist: guiding our steps, inspiring engineers, shaping digital marketplaces. Even failures instruct, reminding us that housing human life at scale is both achievement and hazard.

This chapter approaches cities as active participants in history, not mere settings. They are repositories of collective decisions, embodiments of abstract ideals, and laboratories of resilience and collapse. To understand them is to perceive how the material and the philosophical converge in stone, water, and street. The city is not just where we dwell—it is a statement of what living together might mean.

The Logic of Grids: How City Layouts Embodied Political and Philosophical Order

The straight line is among the simplest of human inventions, yet when multiplied into streets, blocks, and intersections, it becomes a profound statement about how people should live together. The decision to lay out a city in a grid is neither arbitrary nor purely functional: it is an act of political imagination and philosophical assertion. From the Indus Valley to Imperial Rome, from the ideal plans of Greek colonies to the ambitious blueprints of Enlightenment Europe, grids told inhabitants that life within the city was to be shaped by principles larger than personal preference or natural topography. The grid was, and remains, a silent manifesto.

One of the earliest known examples comes from Mohenjo-Daro and Harappa, flourishing in the Indus Valley around 2500 BCE. These cities were not accidental accretions of winding paths but carefully ordered spaces where streets intersected at right angles and drainage systems were standardised across entire neighbourhoods. Such order suggests more than engineering efficiency. It signals a society that conceived of urban life as a shared enterprise, one in which public hygiene and equal access to thoroughfares were not luxuries but essentials. The Indus grid did not simply move carts and pedestrians; it articulated an ethic of civic regularity. To live there was to inhabit a world structured by unseen but pervasive rules.

When the Greeks developed their colonial cities, the logic of the grid was reinterpreted through political philosophy. Hippodamus of Miletus, often credited as the father of urban planning, advocated layouts in which land was divided into equal plots and streets ran in rectilinear fashion. His designs were not mere geometrical abstractions but embodied democratic aspirations: the city as a place where order reflected fairness, where citizens shared equally in its structure. The agora, placed deliberately at the intersection of major axes, reinforced the centrality of public life and debate. In this sense, the Greek grid was not just spatial—it was constitutional. The very plan of the city rehearsed the values of balance, equality, and civic participation. *The geometry of streets was a philosophy in stone, a map of how justice and order might be lived out in daily routines.*

The Romans, inheriting and adapting the Greek approach, elevated the grid into an instrument of empire. The so-called castrum plan, originally for military encampments, with its two main intersecting streets (cardo and decumanus) and neatly divided insulae, became a template for cities across Europe, North Africa, and the Near East. To impose a grid on conquered territory was to imprint Roman order onto foreign landscapes, a tangible expression of control and integration. Unlike the Greek grid, which leaned towards ideals of civic equality, the Roman version was more concerned with authority, logistics, and permanence. Roads aligned with military precision, facilitating troop movement and administration, while the forum at the intersection symbolised Rome's central authority. For the Romans, the grid was empire made visible: it turned local space into imperial system.

Yet grids were never only about power from above. They also shaped daily life from below. The predictability of layout made cities legible to inhabitants and newcomers alike. Navigation became easier; property could be bought, sold, and taxed with greater clarity; public services such as water distribution and waste disposal could be standardised. This rationalisation had a democratising effect: individuals were not at the mercy of labyrinthine alleys or private monopolies over space. Instead, they inhabited a shared design. But predictability also came at a cost. Critics across history have noted that strict grids can be alienating, erasing the organic irregularities that make spaces feel human-scaled and alive. The tightly regimented plan of Roman colonies, for instance, could seem oppressive compared to the winding, adaptive streets of older Mediterranean towns. Thus, *the grid embodies a tension: it offers clarity and equality but risks rigidity and impersonality.*

Nowhere is this tension more visible than in the Enlightenment's rediscovery of the grid as a symbol of rational progress. The Spanish expansion into the Americas exported the Laws of the Indies, prescribing a standard grid with a central plaza, from which streets extended at right angles. These plans sought to impose order on the chaos of conquest, embedding both Catholic and imperial values into the very stones of new cities. The design reflected a theological as well as political claim: that society could and should be organised according to universal rules. Later, in the nineteenth century, Ildefons Cerdà's plan for Barcelona's Eixample reinterpreted the grid yet again, this time as a utopian social project. Cerdà's wide boulevards, chamfered corners, and equal-sized blocks were conceived to alleviate overcrowding, improve sanitation, and promote egalitarian living. His grid was an ethical as much as an architectural project, a direct critique of medieval congestion. Yet, ironically, over time the very regularity he prized facilitated rampant speculation and density, subverting his egalitarian aims.

What becomes evident through these examples is that grids are never neutral. They are spatial answers to questions societies ask themselves: How should power be distributed? What does fairness look like in built form? How should a collective balance efficiency with individuality? To walk a grid is to move through a social contract, whether one recognises it or not. The grid channels not only feet and carts but imaginations and expectations. Its regularity teaches citizens that order is the norm, that life should follow legible rules, and that space is a resource to be divided rationally. This pedagogical role of the grid—its capacity to shape how people think about their world—is perhaps its most enduring power.

Yet grids have always provoked dissenters. Jane Jacobs, in her defence of irregular, mixed-use neighbourhoods in twentieth-century New York, argued that the vitality of a city came precisely from unpredictability, from the mingling of functions and forms that grids often suppress. Others have noted that strict geometries can impose top-down visions that ignore local needs, erasing cultural particularities in favour of abstract uniformity. Indeed, history is full of failed grids—planned towns that never thrived because their imposed order stifled organic life. This raises an important insight: *the grid's success depends on its capacity to negotiate between imposed structure and lived complexity.*

What is striking is how resilient the grid remains. Despite its critics, it continues to dominate the imagination of planners worldwide. Its appeal lies partly in its adaptability: a Roman centuriation scheme and a Manhattan block plan differ in scale and intent, but both embody the same underlying principle of order through repetition. Its endurance also lies in its symbolic power. A grid promises legibility in a world prone to disorder. It assures rulers of control and citizens of predictability. It reduces the chaos of collective life into something measurable, divisible, and graspable.

To study the logic of grids, then, is to confront the ways human beings have sought to domesticate chaos through geometry. It is to recognise that every right angle on a city map is both a technical solution and a philosophical proposition. Ancient planners were not merely drawing lines—they were imagining forms of life. And the traces of their imagination remain beneath our feet, guiding how we walk, where we meet, and what we believe order should look like in the midst of complexity. The grid, far from being a relic of antiquity, continues to remind us that space is never innocent. It is always a mirror of our politics, our philosophies, and our deepest assumptions about what it means to share a world.

Infrastructure as Philosophy: Aqueducts, Sewers, and Roads as Designed Expressions of Collective Priorities

Civilisation is not measured only by its monuments, temples, or palaces but equally by its hidden arteries—the aqueducts that carried water across valleys, the sewers that channelled waste beneath crowded streets, and the roads that connected hinterlands to centres of exchange. These structures are often treated as feats of engineering, impressive for their technical ingenuity. Yet their deeper significance lies in the values they embodied. Infrastructure was not simply practical; it was philosophical. Each aqueduct or road was a statement about what a society deemed essential, about how collective life should be ordered, and about where authority resided. To trace these systems is to uncover the priorities and principles that ancient communities inscribed into stone, brick, and clay.

Consider the Roman aqueducts, arching across landscapes with serene grandeur. They were marvels of hydraulic engineering, yes, but they were also an articulation of a civic vision. Rome, a city of perhaps a million inhabitants at its height, could not have sustained itself without the steady flow of fresh water delivered from distant springs. The decision to invest colossal resources in such works reflected more than a concern for health or convenience; it demonstrated a conviction that urban life deserved permanence and abundance. By guaranteeing plentiful water, the state signalled its role as custodian of collective wellbeing. Aqueducts also served a symbolic function. Their monumental visibility reminded every passer-by of the empire's reach and of its capacity to master nature for public benefit. *The aqueduct, in other words, was an argument in stone: that order, stability, and civic flourishing required the deliberate intervention of collective power.*

Yet this logic was not uniquely Roman. In ancient Mesopotamia, the construction of canals to irrigate fields was a matter of survival in an arid landscape, but the organisation of labour to maintain them carried philosophical weight. The canal system bound communities together, demanding cooperation, oversight, and continuous negotiation between farmers, priests, and rulers. Control of water became synonymous with control of life itself, and the distribution of that water mirrored the distribution of authority. Infrastructure, then, was not neutral. It channelled power as surely as it channelled water, embedding hierarchies into the daily routines of those who tilled the land.

Sewers provide another case where technical necessity converged with philosophical outlook. The Great Drain of Mohenjo-Daro in the Indus Valley, dating back over four millennia, reveals a civilisation that treated sanitation as a public concern. Streets were laid with covered drains that connected to household bathing areas, suggesting that cleanliness was not merely a private virtue but a civic principle. Similarly, Rome's Cloaca Maxima—initially built to drain marshlands and prevent flooding—came to stand as a model of how a city could manage its own by-products, turning waste into a matter of collective governance. Such investments expressed a recognition that the health of individuals depended upon shared infrastructure. Philosophically, this marked a departure from the idea that misfortune was the burden of the household alone. To manage waste publicly was to affirm that society had obligations extending beyond the visible spheres of law and defence. It was an early form of what we might now call social responsibility.

Roads, too, were more than logistical conveniences. Their alignment, construction, and maintenance articulated conceptions of community and authority. The famed Roman network of roads, stretching over 80,000 kilometres at its peak, was designed not only to move legions swiftly but also to integrate a vast and diverse empire. Every milestone, every straight stretch cutting through hills, proclaimed the centrality of Rome. To travel on such roads was to participate in an imagined community, one in which distant provinces were not isolated peripheries but integral components of a greater order. The philosophy here was imperial: connectivity as a mechanism of unity, infrastructure as a visible reminder that disparate peoples shared not only a political overlord but a common material framework.

This intertwining of technical function and civic meaning can be seen in how infrastructure mediated between the natural and human worlds. Aqueducts tamed rivers, sewers channelled refuse, and roads domesticated landscapes. Each represented an assertion that human societies were not passive before nature but capable of shaping it deliberately. At the same time, these projects required collective sacrifice—labour, taxation, and obedience. They remind us that infrastructure is always a negotiation between liberty and order, between the autonomy of the individual and the demands of the collective. *Every bridge or sewer is thus a philosophical text: it describes, in concrete form, how a society conceives of the balance between necessity, fairness, and authority.*

There were, however, tensions embedded in these choices. Large-scale infrastructure could enhance equality by providing universal benefits, but it could just as easily reinforce inequality if access was uneven. Roman aqueducts supplied public fountains and baths, yet elite households often enjoyed private conduits delivering water directly, while poorer districts depended on communal outlets. Roads facilitated trade and communication, but they also expedited conquest and extraction, reminding subject peoples of their subordination. Even sewers, though improving public hygiene, often reflected and reinforced social divides, as waste from elite households was removed more effectively than that from crowded tenements. These contradictions reveal that infrastructure always carries a dual identity: it promises universality while often encoding hierarchy.

The philosophical weight of infrastructure becomes clearer when one contrasts societies that invested heavily in it with those that did not. The Maya, for instance, constructed elaborate ceremonial centres but relatively modest systems of water management compared to Rome or Mesopotamia. This reflects not a deficiency of skill but a different prioritisation: ritual architecture was emphasised over hidden utility. The absence of extensive aqueducts or sewers signalled a conception of collective life in which symbolic and religious expression outweighed concerns for material regularity. Thus, infrastructure—or its neglect—becomes diagnostic of what a society valued most.

Modern parallels underscore the timelessness of this logic. When contemporary cities build underground metros or high-speed rail lines, they are not simply solving transport problems but announcing their aspirations for connectivity, sustainability, or global prestige. When nations neglect crumbling water systems, they are not merely failing at maintenance but revealing what they refuse to prioritise. The same questions faced by Rome or Mohenjo-Daro—who controls resources, who benefits, who sacrifices—persist beneath the asphalt and concrete of today's cities. Infrastructure remains philosophy by other means.

Perhaps the most profound insight to be drawn is that infrastructure teaches citizens how to think about themselves in relation to the collective. To walk on a Roman road was to know oneself as part of an empire; to draw water from an aqueduct-fed fountain was to experience the reach of a state beyond the horizon; to trust in a sewer system was to rely on a community's commitment to shared health. *Infrastructure shaped not only landscapes but consciousness.* It habituated people to certain expectations of order, responsibility, and interdependence. In this sense, it was not simply built environment but moral environment.

Aqueducts, sewers, and roads endure in ruins and replicas because they crystallised enduring truths about human organisation. They remind us that beneath technical ingenuity lies a more profound layer: a society's judgment about what deserves collective effort, what burdens it chooses to share, and what vision of common life it seeks to project into the future. To study these works is to encounter not only ancient engineering but ancient philosophy made durable, cast into stone, and lived in daily routine. They are, quite literally, thought made infrastructure.

Markets and Information: How Spaces of Exchange Became Systems for Circulating Knowledge

A market has never been only a place to buy and sell. To stand in an ancient agora, a medieval bazaar, or a Renaissance piazza was to be immersed in a constant flow of information. Prices revealed the pressures of supply and demand, but words, rumours, observations, and practices carried their own value, diffusing through crowds as surely as coins passed from hand to hand. Long before the age of printing presses or digital networks, markets functioned as hubs of communication where news, techniques, and ideas were traded as eagerly as spices or metals. *The circulation of goods and the circulation of knowledge were inseparable, each reinforcing the other.*

The agora of classical Athens illustrates this dual character vividly. Philosophers such as Socrates and later the Sophists chose the marketplace not for its convenience but because it was the centre of public life. It was there that conversations could reach a wide and varied audience, that ideas could be tested against the responses of merchants, sailors, artisans, and citizens alike. Knowledge did not travel in scrolls alone; it was carried in speech, debate, and observation. The very architecture of the agora, with its open colonnades and gathering spaces, facilitated not only trade but intellectual exchange. One could learn of distant battles from returning hoplites, hear political rumours whispered by rivals, or absorb cultural trends through imported pottery or textiles. *The market was a classroom, newsroom, and parliament all at once, because it condensed the flows of the city into a single, pulsating space.*

In Mesopotamia, marketplaces were equally vital to the transmission of knowledge. Clay tablets unearthed from merchant households reveal not only commercial records but technical instructions, religious texts, and even medical recipes. Trade routes brought lapis lazuli from Afghanistan and tin from Anatolia, but they also brought new accounting practices, foreign gods, and unfamiliar techniques of metalwork. To be a merchant was to be a carrier of culture, not merely of commodities. The contracts they wrote, the standard weights they used, and the calendars by which they marked debts all represented a sophisticated informational infrastructure. Commerce demanded accuracy, but accuracy itself depended on shared conventions—systems of measurement, trust in records, and collective memory. Thus, economic exchange was inseparable from epistemic order.

Markets also served as laboratories of practical knowledge. Consider the medieval fairs of Champagne in France. These gatherings drew merchants from across Europe and the Mediterranean, creating temporary towns where not only goods but also innovations in credit, law, and logistics circulated. Bills of exchange, letters of credit, and early forms of insurance spread through these fairs, transforming the possibilities of long-distance trade. None of these instruments could have flourished in isolation; they required dense human networks where reputation, precedent, and experimentation could be tested. The fairs embodied the principle that *information needs proximity*, that ideas acquire durability not simply by invention but through repeated use, contestation, and adoption by communities of practice.

One might be tempted to imagine markets as purely egalitarian spaces of information-sharing, but the reality was more ambivalent. Knowledge was not evenly distributed. Merchants who travelled widely, who spoke multiple languages, or who had access to literacy enjoyed significant advantages. Markets could democratise access to news—anyone could overhear a rumour—but they could also reinforce hierarchies by allowing those with privileged information to profit disproportionately. This tension remains familiar today, in debates about insider trading or data monopolies. The same dynamics that once allowed a well-informed silk merchant in Venice to outmanoeuvre his rivals underpin the algorithms that today determine who benefits from digital marketplaces. Markets have always been theatres of asymmetry as well as arenas of exchange.

Importantly, markets transmitted not only commercial or political news but embodied knowledge: techniques, styles, and tastes. A new method of weaving, a culinary innovation, or a style of dress could ripple outward through imitation and adaptation. The spice trade exemplifies this phenomenon. Pepper, cinnamon, and cloves did not only flavour European diets; they carried with them knowledge of foreign climates, agricultural practices, and navigational routes. Each sack of spice was a fragment of a distant ecology, a condensed record of what was possible elsewhere. Markets made such fragments visible, tangible, and replicable, allowing cultural knowledge to travel through taste, fashion, and material design.

This role of markets as information systems becomes clearer when one considers their fragility. When plague struck medieval Europe, the interruption of markets not only starved economies of goods but severed channels of communication. Isolation bred ignorance as much as scarcity. By contrast, the reopening of markets after outbreaks often coincided with bursts of innovation, as pent-up ideas and techniques flowed once again across regions. The dependence of knowledge circulation on the maintenance of physical marketplaces reminds us that ideas require infrastructure: places, practices, and rituals where humans can meet, observe, and compare.

The philosophy underlying these exchanges was subtle but profound. To enter a market was to acknowledge a world larger than one's household or village, a world in which strangers could be trusted, if only temporarily, to set terms of value and share fragments of knowledge. Trust was both the precondition and the by-product of exchange. Without it, contracts and bargains would collapse; with it, information could spread beyond kinship lines and local traditions. *Markets taught communities to think beyond the parochial, to orient themselves toward horizons they could not see but could nonetheless imagine through the goods and stories that arrived from afar.*

Modern parallels are instructive. Stock exchanges, with their frenetic trading floors and electronic screens, operate as real-time information markets where prices aggregate countless bits of knowledge dispersed across the globe. Digital platforms such as Amazon or Alibaba extend this logic, transforming the circulation of reviews, recommendations, and consumer data into forms of knowledge as valuable as the products themselves. The principle remains consistent: markets are systems for processing information, compressing dispersed signals into shared reference points. What has changed is the scale, speed, and opacity of the process. Where once a merchant's shouted news might sway the price of wool in a town square, now algorithms digest billions of data points in milliseconds. The informational role of markets has expanded, but the underlying dynamics remain recognisable.

There is a deeper lesson here about the relationship between knowledge and space. Markets were not abstract networks but embodied settings: noisy, crowded, material environments where people watched, listened, and bargained. Information was not simply transmitted but performed—through gestures, displays, rituals of weighing and testing. Even today, the vibrancy of face-to-face markets, from street bazaars in Marrakesh to farmers' markets in London, demonstrates that information thrives when it is embedded in sensory experience. Prices on a screen cannot replace the tacit knowledge gleaned from smell, touch, or tone of voice. *Markets remind us that knowledge is not merely data but lived interaction.*

By studying ancient and medieval markets as systems of information, one begins to see them less as backdrops to history and more as engines of transformation. They were the places where innovations in finance stabilised long-distance trade, where political rumours ignited uprisings, where techniques of craft spread across cultures, and where philosophical debate mingled with commercial bargaining. Their legacy lies not only in the wealth they created but in the webs of understanding they wove, webs that sustained societies and propelled them into new forms of life. Markets were not just sites of economic necessity; they were crucibles of knowledge, shaping how human communities perceived, organised, and expanded their world.

Urban Fragility: Why Ancient Cities Collapsed and What Their Limits Reveal

The rise of cities has often been framed as humanity's most durable achievement, yet history offers an unsettling counterpoint: cities are also among its most fragile creations. For every Rome or Cairo that managed to endure across centuries, there are many more that faltered, dwindled, or disappeared altogether. Ur in Mesopotamia, Teotihuacan in Mesoamerica, Angkor in Cambodia—each once commanded vast hinterlands and radiated cultural influence, only to become husks of former greatness. These collapses were rarely sudden implosions; they were more often the outcome of stresses that revealed the limits of human-scale innovation. Cities concentrated wealth, talent, and ingenuity, but they also amplified vulnerabilities. *Urban fragility was not an aberration of ancient civilisation—it was one of its defining features.*

Consider the case of Ur, once the jewel of Sumerian Mesopotamia. Its prosperity was built on irrigation, a system that demanded extraordinary coordination among thousands of farmers, engineers, and administrators. The canals that carried water also required constant dredging, maintenance, and recalibration. Yet as salinisation accumulated in the soil and political unity fractured, the very system that had allowed Ur to flourish became unsustainable. Farmers abandoned fields, food production declined, and the city's capacity to support its dense population weakened. The innovation of irrigation had expanded human possibility, but it had also created an ecological trap: *a system so effective in the short term that it obscured the longer-term costs.* Cities did not fall because humans lacked ingenuity; they fell because their ingenuity could not always anticipate the unintended consequences of scaling it.

Teotihuacan, which flourished between the first and seventh centuries CE in central Mexico, offers another lesson. It was one of the largest cities of its time, home to more than 100,000 people, with its vast Avenue of the Dead and monumental pyramids testifying to its grandeur. But its decline was marked by internal unrest and eventual destruction from within. Archaeological evidence points to revolts against the elite and the burning of elite compounds. What is striking here is that collapse came not from external invasion but from the fraying of internal cohesion. The very density that had allowed Teotihuacan to generate wealth and culture also magnified inequality and political strain. When trust in leadership broke down, the system could not absorb the shock. This highlights a persistent tension: cities thrive on cooperation among strangers, but that cooperation is fragile, dependent on legitimacy and reciprocity. *Without a shared sense of fairness, complexity can unravel into conflict.*

Environmental shocks often exposed the precariousness of urban systems. Angkor, capital of the Khmer Empire, reached its zenith in the 12th century with its sprawling temple complexes and sophisticated hydraulic infrastructure. The reservoirs and canals that sustained rice agriculture enabled Angkor to dominate its region for centuries. Yet palaeoclimate studies suggest that prolonged droughts followed by intense flooding in the 14th and 15th centuries disrupted this delicate system. The very scale of Angkor's waterworks, once a triumph of human ingenuity, made it inflexible when confronted with climatic volatility.

Smaller communities might have adapted by shifting cultivation patterns; Angkor, locked into monumental infrastructure, could not pivot quickly enough. The city's fragility was a function of its ambition. *The larger and more intricate the system, the less margin it had for improvisation.*

Urban collapse also reveals the paradox of connectivity. Cities existed not as isolated entities but as nodes in wider networks of trade, migration, and cultural exchange. This interdependence enriched them but also made them vulnerable. When Mediterranean trade contracted after the fall of Rome, many urban centres shrank dramatically because their economies had been tied to long-distance flows of goods and tribute. Likewise, cities along the Silk Roads thrived when caravans brought steady traffic, but declined sharply when routes shifted or political disruptions made trade unsafe. The same openness that brought prosperity could, under changing conditions, expose cities to devastating isolation. *Fragility was the shadow side of connectivity.*

One cannot ignore the role of disease in this story. Urban density created opportunities for knowledge, art, and commerce, but it also fostered epidemics. The Antonine Plague in the Roman Empire, likely smallpox, and the later Justinianic Plague, devastated populations and undermined imperial structures. Ancient cities lacked the biomedical knowledge to control pathogens, but more importantly, their very form—the crowding of people, the reliance on granaries, the movement of soldiers and traders—acted as amplifiers of contagion. Cities embodied both the promise of collective life and its most dangerous vulnerabilities. *To build a city was to wager that human organisation could outpace the risks of proximity.*

It would be misleading, however, to treat all collapses as failures in a moral sense. Many ancient cities did not vanish entirely; they contracted, adapted, or transformed into new kinds of settlements. Rome itself, often described as "falling" in the 5th century, continued to exist, albeit on a reduced scale, its population shrinking from over a million to tens of thousands. What collapsed was not the physical city alone but a particular mode of urban life—its imperial infrastructure, its political order, its economic scale. This distinction matters. *Urban fragility is not always about disappearance but about the limits of sustaining certain forms of complexity.*

What, then, does this tell us about the boundaries of human innovation? Ancient cities demonstrate that ingenuity is always entangled with ecology, politics, and social trust. Irrigation, monumental waterworks, or trade networks expanded human capacity, but they also created rigid dependencies. Fragility emerged when systems became too tightly coupled, when their success made them less adaptable, or when inequalities eroded the legitimacy required for cooperation. In this sense, collapse was not simply a technical failure; it was a philosophical reminder of scale. Human innovation thrives within certain bounds, but beyond those bounds, the very features that make cities flourish can accelerate their decline.

The relevance of these lessons is not confined to antiquity. Modern cities, from São Paulo to Shanghai, face challenges of water scarcity, inequality, climate volatility, and disease spread. The difference is scale and speed, not essence. Contemporary urban centres, reliant on global supply chains and digital networks, are even more tightly coupled than their ancient predecessors. Yet the principles remain:

success breeds fragility, complexity requires resilience, and legitimacy underpins cooperation. To study the fragility of Ur, Teotihuacan, or Angkor is not an antiquarian exercise but a mirror held up to our present condition. *The fate of ancient cities shows us that fragility is not an exception to urban life—it is its constant companion.*

Chapter 10
The Transmission of Ingenuity

Ingenuity has seldom respected borders. While empires rose and fell, and kings fought to delineate the limits of their power, ideas moved along routes indifferent to such lines. The spread of mathematics, medicine, and engineering across civilisations was never neat or orderly; it was messy, contested, and transformative. No society created in isolation. Each borrowed, adapted, and reinterpreted knowledge from its neighbours. What travelled was not simply information but understanding itself, continuously reshaped in the process. Transmission was as much about reinvention as preservation.

The Silk Roads exemplify this phenomenon. Often imagined merely as trade routes, carrying silk from China, spices from India, or glass from the Mediterranean, they were equally arteries of intellectual exchange. Caravans transported physicians sharing herbal remedies, mathematicians demonstrating calculation methods, and engineers conveying architectural knowledge. A Persian scholar might learn an Indian computation technique, carry it westward, and teach it to a Syrian student, who would refine it in Arabic. By the time such knowledge reached a European scriptorium centuries later, its origins were obscured, yet its impact remained profound. Consider the story of "Arabic numerals": developed in India, elaborated by Islamic mathematicians, and transported into Europe, they transformed arithmetic, accounting, architecture, and navigation alike. This was not mere transmission—it was transmutation. Ideas rarely arrived intact; they were reshaped to fit new linguistic, cultural, and practical contexts.

Central to this process was translation. Far from a mechanical act of substituting words, it demanded interpretation, selection, and at times invention. The Abbasid Caliphate, particularly the House of Wisdom in ninth-century Baghdad, offers a striking example. Greek, Persian, and Indian texts were rendered into Arabic, but translators like Hunayn ibn Ishaq did more than copy—they interrogated, clarified, and occasionally corrected. The result was not replication but creation. Translators faced choices that altered meaning: Greek terms might lack Arabic equivalents, necessitating new words or phrases. In doing so, they expanded the expressive capacity of the receiving language, shaping not only communication but the very way ideas could be conceived. Translation was a creative act that extended the horizons of thought.

Cross-pollination magnified these effects. Greek astronomy, combined with the computational rigour of Islamic mathematics, yielded more than a sum of parts: it produced novel synthesis. The Ptolemaic system of epicycles underwent rigorous recalculation, generating planetary tables that influenced European astronomers, including Copernicus. Indian advances in trigonometry were absorbed and expanded upon by Islamic scholars, then transmitted westward. The cumulative impact was exponential: ideas born in one culture, refined in another, and disseminated further could catalyse entirely new disciplines. The medieval European revival of science leaned heavily on translations from Arabic into Latin, notably in Toledo, where Jewish, Christian, and Muslim scholars collaborated. What arrived in Europe was not pure

Greek thought restored, but Greek thought refracted through centuries of Islamic engagement. Much of what we call "Western science" is, in truth, a global palimpsest.

Yet the transmission of knowledge was never free of tension. Cultures wrestled with whether foreign ideas undermined or enriched their own traditions. In China, Buddhist texts translated from Sanskrit were initially suspect, perceived as alien to Confucian order. Over time, however, translation reshaped Buddhism and Chinese philosophy alike, producing uniquely Chinese schools of thought. In Europe, Renaissance humanists sometimes bristled at reliance on Arabic intermediaries, even as they absorbed the fruits of Islamic scholarship. The impulse to claim originality, to erase signs of indebtedness, coexisted with the drive to learn. Borrowing and adaptation were entwined, revealing the ambivalence of societies confronted with brilliance originating elsewhere.

This ambivalence underscores a crucial point: ingenuity is not the fixed possession of any civilisation. It circulates, is contested, and is continually reimagined. Algebra illustrates this vividly. Its very name derives from the Arabic "al-jabr," from al-Khwarizmi's ninth-century treatise, yet the discipline drew upon Indian arithmetic and Greek geometry. By the Renaissance, algebra had traversed so many hands that pinpointing a singular origin becomes futile. Ownership mattered less than transformation: each society reimagined the discipline to meet its needs. The same holds true for medicine: Hippocratic texts travelled through Islamic hospitals to European universities, acquiring new methods and shedding old assumptions along the way. Knowledge endured because it changed.

Modern parallels clarify the lesson. The contemporary internet, celebrated as a vehicle of global information exchange, mirrors in digital form what the Silk Roads achieved physically. Ideas move rapidly, yet they are rarely received unchanged. A technology developed in Kenya, such as mobile banking, may be adapted differently in India or Latin America. Local culture, regulation, and needs shape reception. Ancient ingenuity reminds us that breakthroughs are seldom isolated sparks; they are the culmination of centuries of cross-cultural fertilisation.

The transmission of knowledge forces us to reconsider originality. If every idea arises from previous influences, value lies not in absolute novelty but in the ability to adapt, recombine, and reconceive. Translation, synthesis, and even appropriation are central to innovation. This is not merely a historical insight; it challenges contemporary assumptions about intellectual property, cultural ownership, and the politics of invention. Debates over patents, digital piracy, and the ethics of AI datasets echo the same tensions that once surrounded mathematical techniques or sacred texts. Understanding how ingenuity moved in the past illuminates the contested terrain of innovation today.

This chapter traces the diverse pathways by which knowledge travelled, transformed, and occasionally met resistance. From the Silk Roads' bustling exchanges to the painstaking labour of medieval translators, from the synthesis of astronomical traditions to the uneasy negotiations between cultural pride and borrowed brilliance, the narrative is not linear but interactive. What emerges is a vision of ingenuity not as the possession of a single people or era, but as humanity's shared inheritance, endlessly reshaped through contact, translation, and adaptation. To trace the transmission of ideas is to recognise that our most remarkable achievements have always been collaborative ventures across time and space.

Silk Roads of Knowledge

When historians speak of the Silk Roads, they usually evoke caravans bearing bolts of silk, chests of spices, and precious stones moving across deserts and mountains. Yet the most transformative cargo was intangible: techniques, theories, and practices in mathematics, medicine, and engineering that travelled across empires and endured long after the goods had faded. The Silk Roads were not merely trade routes; they were intellectual arteries that stitched together distant civilisations into a shared, though uneven, web of ingenuity. *The mathematics used by European navigators, the medical theories practised in Islamic hospitals, and the engineering principles embedded in Chinese waterworks were each, in different ways, products of this ceaseless exchange.*

The importance of mathematics in this circulation cannot be overstated. Indian scholars had devised advanced numerical concepts, including the positional decimal system and the notion of zero, centuries before these ideas reached Europe. Their transmission occurred not by a single leap but through successive adaptations: translated into Arabic in Baghdad, refined through algebraic treatises, and finally absorbed by European scholars through Latin translations in Spain and Sicily. The very word "algorithm" derives from al-Khwarizmi, a Persian mathematician whose ninth-century works blended Indian numerals with Greek geometry. By the time Italian merchants and scholars adopted these numerical methods in the Renaissance, the system was so naturalised that its hybrid origins were obscured. Yet without this lineage, modern finance, navigation, and science would have been unthinkable. The Silk Roads remind us that what we often consider "European mathematics" is in fact a global inheritance.

Medicine too was a field transformed by these exchanges. Greek medical theory, codified in the writings of Hippocrates and Galen, spread eastward and met Indian and Persian traditions. In Baghdad's House of Wisdom, scholars such as Hunayn ibn Ishaq translated Greek medical texts into Arabic, but the process was never mechanical. They incorporated Persian pharmacology and Indian surgical practices, producing a body of medical knowledge that was both synthetic and innovative. Hospitals across the Islamic world, from Damascus to Cairo, became centres not just of healing but of research, where techniques were tested and refined. From there, medical compendia travelled westward, entering Europe through Spain. The celebrated Canon of Medicine by Ibn Sina (Avicenna), itself a synthesis of Greek, Persian, and Islamic thought, became the standard medical text in European universities for centuries. *Medicine on the Silk Roads was not preserved in amber; it was continually reshaped by cultural encounter, trial, and reinterpretation.*

Engineering followed a similar trajectory. The spread of hydraulic knowledge illustrates this well. Chinese expertise in canal-building and water management, developed to sustain vast agricultural populations, travelled westward, influencing Islamic engineers. Techniques of waterwheel construction, aqueduct design, and urban plumbing crossed cultural boundaries, adapted to local climates and needs. The noria, a type of waterwheel, was refined in Syria and North Africa before influencing European designs. Architecture too bears the traces of Silk Road engineering:

domes, arches, and vaulting systems circulated between East and West, creating hybrid forms that defy simple classification. These innovations did not travel as blueprints alone; they were transmitted through artisans, builders, and travellers, each of whom carried both practical know-how and the philosophical assumptions embedded within their craft.

What makes the Silk Roads so instructive is not just the breadth of knowledge exchanged but the way ideas changed in transit. Knowledge rarely arrived intact; it was filtered through language, culture, and application. Translation played a decisive role here. Rendering a text from Sanskrit into Arabic, or from Arabic into Latin, demanded more than linguistic substitution. New terms had to be coined, concepts reinterpreted, ambiguities resolved. In the process, the receiving culture shaped the knowledge to fit its own intellectual frameworks. Thus, the zero of Indian mathematics became a tool in Arabic algebra before becoming indispensable to European commerce. Likewise, Galenic medicine acquired new explanatory frameworks in Islamic contexts, blending humoral theory with practical pharmacology. *The Silk Roads show that innovation thrives not in isolation but in the friction of encounter.*

Of course, this circulation was neither uniform nor free of resistance. Some rulers and religious authorities viewed foreign knowledge with suspicion, fearing that it would undermine established traditions. In China, certain strands of foreign astronomy were tolerated when they improved calendrical accuracy but resisted when they challenged cosmological orthodoxy. In Europe, the reception of Arabic science was at times selective: algebra was embraced for its practical utility, while other aspects of Islamic philosophy were downplayed or dismissed. Yet even resistance could not fully arrest the flow. The economic and political incentives for adopting effective techniques—be it a more accurate star chart for navigation, a superior medical treatment, or a more efficient irrigation device—were too compelling. Civilisations might cloak borrowed ideas in the language of their own traditions, but the borrowing itself remained unavoidable.

The Silk Roads also illuminate the tension between ownership and adaptation. No culture could claim exclusive rights to the knowledge they carried. When Europeans later spoke of the "rediscovery" of Greek science, they often elided the fact that those texts had been preserved, critiqued, and expanded upon by generations of Arabic-speaking scholars. Similarly, when Islamic scholars spoke of Indian numerals, they recognised both their origin and their transformation within Arabic algebra. The very act of attributing a discovery to a single culture becomes problematic in this context. *The history of the Silk Roads unsettles the notion of pure originality, revealing knowledge as cumulative, hybrid, and endlessly reworked.*

What, then, can the modern reader draw from this history? At its core lies the recognition that the movement of knowledge was not a luxury but a necessity. Empires that closed themselves off to outside ideas stagnated, while those that embraced exchange flourished. This dynamic is as relevant today as it was then. Just as merchants once carried algebraic methods and medical texts across mountains, today digital platforms distribute algorithms, treatments, and design principles across borders. Yet the same dynamics of transformation, resistance, and appropriation persist. A technology designed in one country may be radically altered in another; an idea that begins in one language may acquire new power in translation. The Silk Roads remind us that *innovation is not about safeguarding purity but about cultivating exchange.*

The story of the Silk Roads of Knowledge is thus not a tale of goods passing hands but of minds reshaping one another. Mathematics, medicine, and engineering were not the possessions of single empires but the products of centuries of circulation and adaptation. They travelled in the minds of traders, the texts of translators, and the skills of artisans. Each culture that received them altered them, sometimes marginally, sometimes profoundly, but always in ways that made them newly relevant. If the Silk Roads are remembered as highways of silk and spice, they should be equally remembered as the world's first great highways of thought. *To follow the path of these ideas is to see human ingenuity not as isolated achievement but as the shared enterprise of civilisations in motion.*

Translation as Innovation

Every act of translation is an act of transformation. To carry words from one language into another is never a simple exercise in substitution; it is a negotiation between worldviews, a re-creation of meaning that often alters the very concepts being transmitted. Far from being a passive activity, translation has historically been one of humanity's most inventive practices, shaping not only how knowledge was preserved but how it evolved. *The spread of science, philosophy, and literature across civilisations was made possible not through perfect fidelity to the original but through creative adaptation, interpretation, and even misinterpretation.*

The House of Wisdom in ninth-century Baghdad offers perhaps the clearest example of translation as a crucible of innovation. Scholars there translated Greek texts on philosophy, medicine, astronomy, and mathematics into Arabic, but they did not treat these works as sacred relics. Translators like Hunayn ibn Ishaq brought an interpretative lens to Galen and Aristotle, choosing Arabic terms that carried resonances the originals did not anticipate. Where Greek lacked precise technical vocabulary, translators created new words; where cultural assumptions differed, they rephrased ideas to resonate with Islamic intellectual frameworks. This was not distortion but the creative labour of making texts usable in new contexts. In doing so, they laid the groundwork for advances in algebra, optics, and medicine that were distinctly Islamic, even as they bore Greek roots. *To translate was to invent the conditions under which knowledge could flourish anew.*

Consider the case of Aristotle. When his works travelled from Greek into Syriac, then into Arabic, and later into Latin, the chain of translations produced shifts in meaning that transformed European intellectual life. The Latin reception of Aristotle in the twelfth and thirteenth centuries was not of "pure" Greek philosophy but of Aristotle refracted through Islamic interpreters like Ibn Rushd (Averroes). His commentaries, themselves shaped by centuries of translation and reinterpretation, introduced Europeans to a version of Aristotle that emphasised rational inquiry in ways that suited scholastic debates. What European universities studied as Aristotelian philosophy was, in fact, a layered construct born of translation. *Translation here was less a mirror than a prism, bending and refracting light into new colours that no single culture alone had produced.*

The story is similar in mathematics. Indian numerals, transmitted to the Islamic world, were not received as immutable signs. Translators had to create an Arabic lexicon for positional notation and the concept of zero, which had no direct equivalent. Al-Khwarizmi's works were themselves translations and adaptations, merging Indian arithmetic with Greek geometry. When these works reached Europe in Latin, they gave rise to what came to be known as "Arabic numerals" despite their Indian origin. Each act of translation involved rethinking how numbers were conceptualised, taught, and used. The system that today underpins global finance and science is not merely Indian or Arabic but the product of centuries of translational innovation.

Translation also brought cultural friction to the surface. When Buddhist texts moved from Sanskrit into Chinese, translators faced immense challenges in reconciling concepts. Sanskrit terms like *nirvana* and *dharma* carried philosophical connotations foreign to the Chinese worldview. Translators had to decide whether to invent new words or borrow Taoist and Confucian vocabulary, thereby altering meaning in ways that profoundly shaped Chinese Buddhism. The resulting hybrid tradition, distinct from its Indian origins, testifies to how translation is as much about cultural negotiation as linguistic equivalence. In this sense, Chinese Buddhism is itself a translation—not of texts alone, but of an entire system of thought remade to fit new intellectual soil.

The creative power of translation also extended to medicine. When Persian scholars translated Indian Ayurvedic texts into Arabic, they did not simply replicate them. They incorporated local pharmacological knowledge, adapting recipes to ingredients available in the Middle East. Ibn Sina's *Canon of Medicine* was therefore not a direct transmission but a synthesis that combined Greek humoral theory, Indian surgical knowledge, and Persian herbal remedies. Its subsequent translation into Latin carried these layered innovations into Europe, where it shaped medical practice for centuries. The text exemplifies how translation can accumulate layers of interpretation, making each version a fresh act of authorship rather than a derivative copy.

Modern scholarship often debates whether mistranslations should be seen as corruptions or as creative leaps. The Silk Roads are filled with examples where "errors" in translation yielded fertile outcomes. When Ptolemaic astronomy entered the Islamic world, translators sometimes rendered Greek terms imprecisely, leading scholars to rework cosmological models. Rather than freezing progress, these imprecisions spurred reinterpretation, pushing astronomy in new directions. In Europe, mistranslations of Aristotle fed debates on the nature of the soul, fuelling theological and philosophical innovation. *What appears as a loss in fidelity is often a gain in intellectual possibility.*

Translation as innovation also forces us to reconsider questions of ownership. Whose knowledge is it when a Greek text is translated into Arabic, expanded upon by Persian scholars, and then translated again into Latin? Each stage introduces new layers of authorship. To attribute the knowledge solely to its "original" author overlooks the creative labour of those who made it legible to new cultures. In truth, intellectual history is a palimpsest, with each translator writing over the traces of predecessors while leaving their own mark for successors to build upon. This is why the concept of "borrowed genius" is misleading; what we call borrowing is often reinvention under another name.

Even in contemporary contexts, translation remains a generative force. The global dominance of English in scientific publishing may conceal the process, but research translated from Chinese, Japanese, or German into English still alters meaning in subtle ways, shaping how discoveries are understood internationally. In technology, the translation of programming concepts across languages and platforms parallels the ancient transmission of mathematics: ideas must be reshaped to function in new systems. And in diplomacy, the choice of one word over another in treaty translations can reshape international relations. Translation continues to be invention, not preservation.

The history of translation as innovation challenges the comforting myth that knowledge moves across cultures unchanged. It demonstrates instead that knowledge only survives by being remade, often in ways its originators could never have foreseen. Translators are not custodians at the edges of history but agents at its centre, architects of the intellectual worlds that followed. To study them is to understand that the most important innovations are not always those that appear entirely new, but those that transform what already exists into something different, usable, and alive. *Translation is thus one of the oldest and most enduring technologies of human creativity—a way of thinking that proves as inventive as any discovery it helped transmit.*

Cross-Pollination of Ideas

Human progress has rarely been the product of isolated genius. The most transformative leaps often occur not when a civilisation invents something entirely new, but when distinct traditions collide, interact, and reshape each other's knowledge. Nowhere is this more vividly illustrated than in the intellectual exchanges that took place along the old trade and scholarly routes that linked Greece, India, Persia, and the Islamic world. *Ideas do not merely travel; they adapt, mutate, and hybridise in contact with new contexts, producing forms of knowledge that no single culture could have generated alone.*

Greek astronomy provides one of the most striking case studies. The models of the heavens articulated by Ptolemy in the *Almagest*—a text that attempted to describe the complex motions of celestial bodies through epicycles and deferents—were transmitted into the Islamic world through Syriac and Arabic translations. Yet the adoption of Greek astronomy was far from passive. Muslim scholars approached these inherited models with the critical tools of their own mathematical traditions, particularly in geometry and trigonometry, which had roots in both Greek and Indian practices. Figures like al-Battani refined planetary models using more precise observations and introduced systematic trigonometric methods to replace cumbersome chord tables inherited from Greek tradition. By reconciling Greek conceptual frameworks with more advanced computational tools, Islamic astronomers generated predictions that were not merely more accurate but more mathematically elegant. *It was this synthesis—Greek structural ambition merged with Islamic numerical innovation—that set the stage for later European breakthroughs in astronomy.*

This process of cross-pollination was not confined to astronomy. In mathematics, the fusion of Indian numeral systems with Greek geometrical reasoning through Arabic intermediaries produced a toolkit that transformed calculation. Al-Khwarizmi's treatises drew upon Indian arithmetic while structuring them within the logical style of Greek mathematical exposition. The result was not simply a translation of Indian methods but the creation of algebra as a discipline—a way of reasoning that could generalise numerical problems beyond specific cases. When these texts entered Europe through Latin translations in the twelfth century, they carried with them a form of mathematics born of cultural hybridity. The so-called "Arabic numerals" that underpin modern science and finance were themselves products of this complex chain of borrowing, modification, and integration.

Medicine, too, illustrates how intellectual hybridisation created knowledge that was greater than the sum of its parts. Galen's medical theories, firmly rooted in Greek humoralism, were preserved in translation but critically reinterpreted by physicians in the Islamic world. Drawing on Persian pharmacology and Indian surgical practices, scholars like Ibn Sina crafted texts that integrated and expanded upon their sources. The *Canon of Medicine* became not merely a repository of Greek learning but a comprehensive medical system that guided practice across the Islamic world and later medieval Europe. Its authority derived not from fidelity to any single tradition but from the creative weaving together of multiple strands of knowledge into a coherent, adaptable framework.

What is most striking about these examples is the willingness of scholars to treat inherited knowledge as provisional rather than fixed. Greek astronomy, though revered, was subject to revision in light of new observations. Indian numerals were not merely adopted but reconfigured into algebraic methods. Galen's doctrines were not recited verbatim but reinterpreted in light of local needs and alternative traditions. *The act of borrowing was inseparable from the act of rethinking.* Knowledge was not transmitted like a sealed parcel but remade in the very process of exchange.

Such hybridity carried tensions. Some thinkers expressed suspicion towards foreign ideas, fearing that they diluted cultural or religious identity. The transmission of Greek philosophy into the Islamic world, for instance, sparked debates over its compatibility with religious doctrine. Scholars like al-Ghazali criticised the overreliance on Hellenistic reasoning, while others, such as Ibn Rushd, defended the integration of Aristotle as essential to rational inquiry. These debates themselves shaped intellectual life, forcing sharper articulations of what could be accepted, what had to be adapted, and what should be rejected. Cross-pollination was never smooth or uncontested; it was a site of negotiation where intellectual identities were forged through selective appropriation.

The long-term effects of this intellectual cross-pollination are difficult to overstate. When Greek astronomy, filtered and refined through Islamic calculation, reached Europe, it became the foundation upon which Copernicus, Galileo, and Kepler built their revolutionary models. When algebra and trigonometry reached European scholars, they unlocked ways of thinking that would underpin modern physics and engineering. Even the medical system that dominated European universities until the seventeenth century was not simply Galenic but a Galen filtered through centuries of Islamic reinterpretation. The intellectual renaissance of Europe cannot be disentangled from the cross-cultural currents that preceded it.

Modern analogies highlight the enduring relevance of this phenomenon. In contemporary science, breakthroughs often arise not from isolated laboratories but from interdisciplinary exchanges—where biology meets computer science, or neuroscience meets philosophy. The globalised nature of research ensures that ideas cross linguistic and cultural boundaries constantly, raising the same challenges of translation, adaptation, and integration faced by scholars a millennium ago. Artificial intelligence research, for instance, borrows as much from statistical traditions as from neuroscience, creating hybrid models that no single discipline could claim as its own. The cross-pollination of ideas remains one of the primary engines of discovery.

What ancient examples remind us, however, is that such cross-pollination requires intellectual humility. It demands a recognition that one's own tradition does not hold a monopoly on truth, and that innovation often requires the courage to borrow, transform, and even discard elements of received wisdom. Civilisations that embraced this principle—Baghdad during the Abbasid era, Córdoba in medieval Spain, Renaissance Italy—became centres of flourishing intellectual life. Those that turned inward and rejected foreign influences often stagnated. *The willingness to allow ideas to mingle, clash, and evolve proved more decisive than the possession of any single piece of knowledge.*

The story of Greek astronomy merging with Islamic calculation is not a footnote in the history of science but a central demonstration of how human ingenuity works. Knowledge rarely advances in a straight line; it leaps forward through encounters across boundaries, through acts of creative reconfiguration that generate something new out of the old. To study these cases is to see that the history of ideas is less about invention in isolation and more about the fertile interactions that occur when cultures meet. The true legacy of these encounters lies not in the preservation of past brilliance but in the transformation of that brilliance into forms that could carry humanity further.

Intellectual Theft and Adaptation

The history of knowledge is often written as a tale of invention, yet much of what we call invention has been forged in the shadows of borrowing. Civilisations rarely admit that their proudest achievements were shaped by the ideas of others. Instead, they veil these borrowings in the language of originality, or, when pressed, dismiss their sources as primitive sparks before the true fire of their own genius. *The tension between cultural pride and borrowed intellect has been one of the most persistent undercurrents in human history, shaping not only the content of knowledge but also the way societies narrate their own intellectual heritage.*

Consider Rome, a civilisation celebrated for its legal system, engineering, and governance, yet one whose intellectual backbone leaned heavily on Greece. Roman elites openly acknowledged their debt to Greek philosophy and literature, but simultaneously recast Greek thought in the service of Roman identity. Cicero's writings on ethics and governance freely drew on Stoicism and the Academy, yet he framed them as tools for shaping Rome's destiny rather than as tributaries of Hellenic brilliance. This selective appropriation allowed Rome to present itself as a civilisation that refined rather than copied. *The act of borrowing was disguised as the act of perfecting.*

Centuries later, medieval Europe adopted a similar posture when confronted with the immense body of knowledge transmitted from the Islamic world. The universities of Paris and Bologna flourished on translations of Aristotle, Galen, and Ptolemy, yet these texts often reached Europe not from Greece directly but through Arabic intermediaries. Latin translations of Ibn Sina's *Canon of Medicine* or Ibn Rushd's commentaries on Aristotle were not simply pass-throughs; they bore the marks of centuries of debate, reinterpretation, and modification within the Islamic intellectual tradition. And yet, when this knowledge became embedded in European learning, its Islamic lineage was frequently downplayed or erased. The Renaissance narrative positioned Europe as the rightful heir of Greece, leaping across centuries to minimise the debt to the scholars of Baghdad, Córdoba, and Cairo. The pride of cultural identity demanded a story of rebirth rather than one of borrowing.

The accusation of theft is itself historically contingent. When the Abbasid caliphs of Baghdad launched the translation movement in the eighth and ninth centuries, importing Greek, Persian, and Indian texts into Arabic, they were not ashamed of this wholesale borrowing. On the contrary, they saw it as a mark of power to absorb the wisdom of conquered or neighbouring peoples. Translators such as Hunayn ibn Ishaq did not merely carry Greek medicine into Arabic; they reshaped terminology, corrected errors, and in the process produced works that were both faithful and innovative. For the Abbasids, the ability to draw upon others was proof of their cultural dominance. By contrast, when medieval Europeans inherited these same texts, pride demanded that they reframe the lineage as a direct inheritance from antiquity, lest the idea of dependence on Muslim scholars undermine their sense of civilisational superiority. *Borrowing is not inherently shameful, but whether it is cast as theft or as mastery depends on who is telling the story.*

Intellectual theft, then, is not merely about plagiarism in the modern sense; it is about the cultural politics of knowledge. When Chinese gunpowder technology spread westwards, European chroniclers frequently reimagined its development as their own refinement rather than an adoption. The compass, too, was claimed as a European advancement despite its Chinese origins. The process is familiar: borrow, adapt, erase the trail, and recast the outcome as invention. Nations and cultures build their prestige on the belief that genius springs from within, even when it is nurtured by seeds from elsewhere.

This raises an uncomfortable but essential point: *adaptation is itself a form of invention.* To translate a text, to recalibrate a technology, or to absorb a concept into a new cultural framework requires creativity. The zero borrowed from Indian mathematicians became something new when Islamic scholars integrated it into positional notation, and something newer still when European merchants and bankers used it to transform trade and finance. Was this theft, or was it innovation? The answer depends on whether one privileges origins or outcomes.

Yet there is a danger in disguising adaptation as originality. When cultural pride suppresses acknowledgment of borrowed genius, it narrows the space for dialogue and obscures the collaborative nature of human progress. The myth of solitary genius—whether embodied in figures like Newton, da Vinci, or Edison—has long obscured the vast networks of borrowing and adaptation that enabled their work. Newton himself famously declared that he stood "on the shoulders of giants," yet even this admission frames his predecessors as scaffolding rather than as equal partners in a continuum of discovery. Intellectual pride makes borrowing invisible, which in turn encourages each generation to repeat the conceit of originality.

The politics of intellectual theft are not confined to the past. In the twenty-first century, debates over cultural appropriation, patent law, and intellectual property echo the same tensions. When pharmaceutical companies develop drugs based on indigenous botanical knowledge, questions arise over whether this constitutes theft or adaptation. When artificial intelligence systems trained on vast datasets replicate artistic styles or human voices, the issue of ownership becomes fraught. Is the innovation in the algorithm, or is it in the collective human input from which it draws? The very fact that such debates persist highlights how uncomfortable societies remain with acknowledging that borrowed genius is often the essence of innovation.

Acknowledging borrowing does not diminish achievement; it enhances it. To recognise that European science was shaped by Arabic translations, or that Roman philosophy rearticulated Greek ideas, is not to accuse these traditions of weakness but to appreciate their strength in adaptation. The ability to take what is foreign, to reshape it until it feels native, and to extend its applications beyond its original context—that is as creative an act as invention in the narrow sense. *Borrowing without transformation is mimicry; borrowing with transformation is the essence of progress.*

The history of intellectual theft and adaptation is ultimately a history of human pride. Cultures want to see themselves as creators, not borrowers. Yet the more honestly we examine the record, the more we see that creation and borrowing are inseparable. Knowledge does not move in sealed containers; it spills, it seeps, it is taken up and reworked until its origins blur. Civilisations thrive not by denying this, but by embracing the generative tension between pride and borrowed genius. It is in this tension—never fully resolved, always contested—that some of humanity's greatest achievements have been born.

Chapter 11
When Innovation Fails

Civilisation's narrative is often told as a series of triumphs—arches that defy gravity, canals that redirect rivers, algorithms that reshape thought itself. Yet entwined with these achievements are quieter, frequently overlooked episodes of failure. These are no less instructive, for they reveal that innovation is rarely a straight ascent. For every dome that still inspires awe, there is a toppled monument, a reminder of ambition untethered from reality. For every engineered landscape that brought abundance, there are fields turned to dust, forests erased from memory. Innovation falters when ingenuity outruns prudence, when invention exceeds its foundations, or when societies forget that technology is a fragile bridge between human intent and natural limits. This chapter explores that precarious tension.

The allure of permanence has long seduced rulers and builders. Stone, unlike flesh, resists decay. To carve it, to raise it skywards, to bend it into unprecedented forms—these were declarations of power, intended to outlive their creators. Yet history's rubble demonstrates that even stone can betray its makers. The Tower of Babel may linger only as allegory, but Mesopotamian ziggurats and Angkor's collapsed temples are tangible testimony to ambition meeting material fatigue. Teotihuacan, in central Mexico, planned with cosmic precision, thrived for centuries before succumbing to ecological and political pressures. Its pyramids remain as stone shells, while the society that built them vanished. Rome's Colosseum survives, in part, because much of it fell; its fractures and missing stones remind us that permanence was always provisional. Hubris in stone is not merely about collapsed walls; it is the arrogance of assuming that matter can bear more than it can, whether structurally, ecologically, or symbolically.

If stone tells of physical overreach, soil tells of ecological oversight. Civilisations depend on sustenance, and sustenance depends on land and water. The Sumerians' first irrigation canals along the Tigris and Euphrates secured extraordinary harvests—but over centuries, salinisation poisoned their fields, rendering once-fertile land unusable. Across continents, similar patterns emerge: the Maya, whose deforestation and relentless cultivation undermined water systems; Ethiopia, where centuries of erosion gnawed at the land that fed its people. Such cases reveal that innovation ignoring natural feedback can rebound catastrophically. Agriculture, engineering, and hydrology brought surpluses—but when abundance encouraged expansion without restraint, the land itself exacted a price.

War deepens the irony. Weaponry has long driven rapid and relentless innovation. Each new advance—from chariots to crossbows, from gunpowder to nuclear fission—was justified as security or dominance. Yet these same innovations often destabilised the societies that produced them. Bronze-tipped spears reshaped battlefields while triggering arms races that bankrupted city-states. The Greek hoplite phalanx shifted Mediterranean power yet empowered ordinary citizens, prompting political fractures at home. Gunpowder transformed not just warfare but state structures, centralising resources around those capable of financing foundries and fortresses. The nuclear bomb epitomises war's paradox: its immense power deters large-scale conflict even as it traps humanity in constant anxiety.

The sharper the weapon, the more peril it carries for its wielder.

Failure is not always spectacular. Sometimes it unfolds invisibly, through forgetting. Knowledge relies on transmission—oral traditions, manuscripts, apprenticeships, institutions. When those systems collapse, wisdom can vanish within a generation. The burning of Alexandria's Library serves as an emblem of lost knowledge, but quieter erasures are equally consequential. When the Western Roman Empire fell, technical expertise in concrete, aqueduct maintenance, and road-building largely disappeared from Europe for centuries. In Mesoamerica, Spanish conquest obliterated not only political order but systems of record-keeping, such as the Inca quipus. Material remnants survived, yet the interpretive frameworks perished with their custodians. Forgetting is collapse in slow motion: a failure of continuity rather than of material strength.

These failures share a common lesson: innovation is not a linear march from ignorance to enlightenment. It is a cycle of trial, error, adaptation, and occasionally irreversible collapse. Collapsed monuments, exhausted soils, runaway weaponry, vanished knowledge—all illustrate that innovation cannot be divorced from humility, foresight, and responsibility. Modern societies are not exempt. Skyscrapers, dams, and megacities testify to ongoing hubris; climate change, biodiversity loss, and depleted aquifers echo ancient ecological miscalculations on a global scale. Cyberweapons and artificial intelligence mirror the paradox of armaments: innovations intended to control may destabilise their creators. Even digital knowledge is fragile, stored on impermanent servers and proprietary formats that may one day vanish as completely as lost scrolls.

Recognising the failure of innovation does not diminish human ingenuity; it illuminates its full complexity. It acknowledges that the same creative impulse that builds can blind, that invention carries seeds of overreach, that pride without restraint courts collapse. Past failures are mirrors, not relics. They compel us to ask whether innovation is sustainable, responsible, and resilient against natural, political, and human pressures. By examining these failures, this chapter does not merely recount mistakes—it invites reflection on the conditions under which human creativity genuinely serves, and when it instead undermines, the foundations it was meant to secure.

Hubris in Stone: Failed Monuments and Collapsed Structures as Lessons in Overreach

Civilisations have always sought permanence through stone. To raise a monument was not simply to build but to proclaim: *we were here, and we mattered*. Yet the story of architecture is as much about collapse as it is about endurance. Across continents and millennia, ruined towers, fractured domes, and half-finished megastructures remind us that ambition too easily slips into hubris. These failures were not accidents of history; they were deliberate acts of overreach, where desire for scale, grandeur, or eternal legacy outstripped the realities of physics, materials, or the societies that commissioned them. To study them is to recognise a recurring human pattern: the belief that ingenuity can outpace nature, and that authority can inscribe itself onto eternity through stone.

Consider the great ziggurats of Mesopotamia, vast terraced structures that once commanded the skylines of Ur and Babylon. Designed as stairways between the heavens and earth, they were built from mudbrick, a material abundant yet vulnerable. The outer layers, often faced with baked brick, shielded the structure for centuries, but over time erosion gnawed at the unprotected cores. With the collapse of the empires that maintained them, these structures decayed into mounds of rubble. The lesson here is sobering: *monumental ambition requires not just construction but continuous stewardship*. Without the societies to sustain them, even the most imposing structures surrender to wind and rain. The hubris was not only in scale but in the presumption that permanence was guaranteed by mass alone.

In medieval Cambodia, Angkor Wat and its surrounding temples stand as both triumph and warning. The Khmer Empire engineered one of the most complex hydraulic systems in pre-modern history, a network of canals and reservoirs designed to regulate water for rice cultivation and to sustain monumental construction. Yet that very system may have contributed to the empire's decline. Deforestation, soil erosion, and over-engineering destabilised the delicate balance between human design and ecological reality. Temples that once projected divine kingship now sit strangled by tree roots, half-consumed by the jungle. The grandeur of Angkor was not undone by invading armies alone but by the ecological fragility that its own ambition unleashed.

Closer to modernity, the leaning tower of Pisa offers another instructive tale. Conceived as a freestanding bell tower to demonstrate civic pride, its foundation was laid upon soft, unstable soil. Almost from the start, the tower began to tilt. Efforts to correct its lean only compounded the problem. For centuries it stood as an embarrassment as much as an icon, a visible reminder that civic pride had overlooked geology. Only in the twentieth century did engineers stabilise the tower, ensuring it would not collapse. What might have been remembered as a failed monument was transformed, through hindsight and modern engineering, into a global landmark. The tower's survival is less a testament to medieval builders than to the humility of later engineers who accepted the flaw and worked within its limits rather than against them.

The fate of grand monuments often mirrored the fragility of the regimes that sponsored them. The Pharaoh Akhenaten's new city at Amarna, constructed with remarkable speed in the fourteenth century BCE, was designed to embody a religious revolution. Yet his reign was short, and his successors rejected his radical theology. The city was abandoned, its temples dismantled, its monuments erased. Here, stone did not crumble through physics but through politics: *a monument is only as eternal as the consensus that sustains its meaning*. Akhenaten believed he had carved his vision into permanence, yet within decades it was rubble, proof that hubris can lie in assuming cultural authority is as enduring as granite.

There is also the matter of sheer technical ambition exceeding practical limits. The Tower of Babel, though mythic, resonates precisely because it illustrates a real pattern: the desire to build beyond human scale. Throughout history, rulers have commissioned projects whose scale outpaced both resources and technique. In Renaissance Florence, for instance, the construction of Santa

Maria del Fiore's vast dome pushed architecture to its limits. It succeeded only because Filippo Brunelleschi devised ingenious methods to overcome the absence of scaffolding tall enough to span its interior. Contrast this with the numerous cathedrals and basilicas across Europe left unfinished, their spires halted mid-construction when funding ran dry or structural integrity proved elusive. These skeletons of ambition speak to the limits of vision unaccompanied by pragmatism.

Philosophically, these failures point to a central paradox of human creativity. Innovation in stone is often born of competition—cities striving to outdo rivals, rulers seeking to impress subjects, empires projecting divine favour. Yet the very impulse that generates advancement also produces overreach. To build taller, heavier, or more intricate is to test not only materials but the endurance of social systems. Skilled labour must be trained and maintained, resources must be extracted, and political stability must persist long enough for completion. When any one of these falters, monuments falter too. *Hubris in stone, therefore, is less about architecture and more about systems thinking: the inability to see that structures are never just physical, but social, ecological, and political assemblages.*

Modern parallels underscore the timelessness of this lesson. The Ryugyong Hotel in Pyongyang, a 105-storey pyramid begun in the late 1980s, was intended as a symbol of North Korea's progress. Poor construction quality, economic crisis, and international isolation left it unfinished for decades, an empty shell dominating the skyline. Even after cosmetic efforts to clad it in glass, it remains largely vacant. Like the ziggurats or Angkor's temples, it embodies the disjunction between ambition and reality. Yet unlike ancient ruins, it has not been softened by centuries of erosion; it remains raw evidence of overreach in the modern age.

Equally instructive are projects that collapse not physically but symbolically. The massive Stalinist architecture of the Soviet Union, from the planned Palace of the Soviets in Moscow to the utilitarian apartment blocks that still line post-Soviet cities, was conceived as enduring proof of ideological supremacy. Yet the palace was never built, abandoned when war intervened, and the blocks that remain now symbolise decline rather than triumph. The material persists, but the meaning has collapsed. This reveals another dimension of failure: *stone can endure, but its cultural significance is far more fragile.*

What unites these examples is a recurring blindness to limitation. Builders underestimated natural forces, overestimated political longevity, or ignored ecological balance. At times they placed too much faith in materials, believing mass could compensate for poor foundations. At other times they assumed societies could sustain projects that strained labour and resources beyond capacity. In each case, the failure of stone became a public lesson in humility, even if that lesson was learned centuries later by archaeologists rather than by contemporaries.

Today's world continues to wrestle with the same temptations. Megacities construct glass-and-steel skyscrapers in seismic zones, betting on engineering solutions that may or may not withstand the long-term tests of earth and climate. Nations build artificial islands, dam colossal rivers, or raise gargantuan statues to project power, each act echoing the ambitions of past rulers who believed their monuments would be unassailable. Yet climate change, shifting political orders, and resource scarcity remind us that permanence remains an illusion. *The ruins of the past are not only archaeological curiosities; they are cautionary mirrors in which we can glimpse our own fragilities.*

To study hubris in stone is not to dismiss human creativity but to temper it with humility. The desire to build will not vanish, nor should it; architecture is one of humanity's most visible expressions of imagination. But when stone fails, it teaches that imagination must be tethered to reality. Enduring structures are those that negotiate with the forces of nature, balance ambition with maintenance, and align with the societies that sustain them. Those that ignore these principles may rise briefly, but eventually, like so many monuments before them, they become silent ruins testifying not to power achieved but to lessons ignored.

Ecological Blind Spots: How Over-Irrigation, Deforestation, and Overfarming Undermined Societies

Civilisations often fall not because of conquest or catastrophe but because they eroded the very environments that sustained them. Human ingenuity in agriculture and water management has repeatedly secured abundance, yet history shows that the same practices, pushed beyond ecological limits, have brought collapse. The pattern is striking: strategies that ensured survival in one generation often sowed the seeds of decline in the next. What undid great societies was not a lack of knowledge but a failure of foresight—a blindness to ecological thresholds that once crossed, could not be easily reversed.

The Mesopotamian plains, often celebrated as the "cradle of civilisation," illustrate this paradox. Early irrigation systems allowed cities like Uruk and Babylon to flourish in otherwise arid landscapes. Yet centuries of channeling water from the Tigris and Euphrates gradually salted the soils. Evaporation left mineral deposits behind, rendering fields infertile. Records from cuneiform tablets chart declining crop yields, and archaeologists have found abandoned settlements where once-fertile land became barren white crusts. The Sumerians had achieved remarkable mastery over their rivers, but their very ingenuity locked them into a system that intensified soil degradation. *Over-irrigation, pursued to maximise production, ultimately undermined the basis of subsistence.*

Deforestation presents a similarly recurrent theme. The rise of the Mayan civilisation in Mesoamerica was inseparable from the expansion of agriculture and monumental architecture, both of which demanded vast quantities of timber. Trees were felled not only for construction but also for lime production—burning limestone to create plaster consumed enormous amounts of wood. As forests retreated, rainfall patterns shifted and soils eroded. Pollen records from lakebeds reveal a sharp decline in tree cover during the Late Classic period, coinciding with political turmoil and urban abandonment. While warfare and dynastic struggles hastened the collapse, environmental fragility amplified the crisis. The Maya did not lack knowledge of agriculture; they practised sophisticated terracing and water storage. But in their pursuit of grandeur, they overlooked the cumulative consequences of deforestation.

In North Africa, the once-prosperous Carthaginian and later Roman provinces relied on intensive farming to supply grain to growing empires. Accounts suggest that regions of modern Tunisia, Algeria, and Libya were highly productive. Yet centuries of overfarming exhausted soils, while the removal of vegetation for pasture and fuel accelerated desertification. The Sahara expanded, and once-thriving breadbaskets declined irreversibly. The ecological changes outlasted the empires themselves, leaving landscapes scarred long after political power shifted elsewhere. *The lesson is that empires can vanish, but ecological damage endures.*

Overfarming is perhaps most clearly illustrated by the American Dust Bowl of the 1930s—a modern echo of ancient patterns. Encouraged by government policy and driven by market demand, farmers ploughed vast areas of the Great Plains, replacing native grasses with wheat monocultures. When drought struck, the exposed topsoil turned to dust, carried across states in storms that blackened skies and displaced millions. This was not ignorance of ecology so much as wilful neglect: local farmers and scientists had long warned of the fragility of the plains ecosystem, yet the pressure of short-term gains silenced caution. The Dust Bowl demonstrates that ecological blind spots persist even in technologically advanced societies.

One of the striking features of ecological mismanagement is its delayed visibility. Over-irrigation does not ruin fields in a single season; salinisation creeps over decades. Deforestation may not collapse rainfall cycles immediately but gradually weakens ecosystems until resilience is lost. Overfarming can sustain yields for generations before soils erode past recovery. This delay fosters complacency.

Leaders and communities often assume stability because crises unfold too slowly to command immediate attention. By the time collapse becomes undeniable, the underlying damage is usually irreversible. *Ecological systems are not infinitely forgiving; they contain thresholds that once crossed, redefine entire landscapes.*

There is also a political dimension to ecological blind spots. Agricultural intensification often benefits elites while shifting costs onto future generations or marginal communities. In Mesopotamia, rulers celebrated the wealth generated by irrigation, even as salinity levels rose. In colonial North Africa, extraction of grain served distant metropoles while degrading local soils. In the Dust Bowl, financial pressures and land speculation encouraged unsustainable practices. The blindness was not solely ecological but moral: decisions were shaped by short-term power, profit, or prestige, rather than by stewardship of the land. Thus ecological failure is rarely accidental; it is entangled with social priorities and political choices.

Yet not all societies succumbed without adaptation. Ancient farmers in the Andes developed terraced agriculture that conserved water and soil on steep slopes, sustaining high-altitude communities for centuries. In parts of Southeast Asia, rice paddies integrated with fish farming created resilient agroecosystems that endured over long timescales. These examples remind us that the problem is not human innovation itself but *innovation without ecological feedback.* Where societies integrated their practices with natural cycles, they sustained prosperity; where they pressed too far against environmental limits, they invited decline.

Modern parallels are difficult to ignore. Large-scale irrigation projects in places like the Aral Sea basin or the Indus Valley have replicated the problems of Mesopotamia, with water diversion leading to salinisation, ecological collapse, and social dislocation. Tropical deforestation in the Amazon echoes the Maya, as trees are cleared for cattle and soy, destabilising rainfall systems not only locally but globally. Industrial farming reliant on monocultures and chemical fertilisers risks creating new Dust Bowls on a planetary scale. Despite centuries of accumulated knowledge, societies continue to repeat familiar mistakes under the pressure of economic growth and political expedience.

What history reveals, then, is not that humans are incapable of sustainable management, but that *blindness arises when societies mistake abundance for permanence*. When fields yield generously, forests appear inexhaustible, or rivers run steadily, it is easy to assume such conditions will continue indefinitely. The hard truth is that abundance is often a fragile state, maintained by balance rather than by human control. Once balance tips, recovery may take centuries or prove impossible altogether.

The enduring lesson is that ecological failure is not a peripheral issue in history—it is central. Civilisations rise not simply by conquering territory or innovating technologies but by managing environments well enough to sustain growth. Their decline is often less about external enemies than about the silent erosion beneath their feet, the gradual withering of soils, forests, and waters. Hubris in stone may produce ruins that stand as visible warnings, but ecological blind spots produce silences—empty fields, vanished forests, desiccated lakes—that are harder to interpret but no less telling.

To grasp this pattern is to recognise that our own civilisation is no exception. Climate change, biodiversity loss, and soil degradation are not abstract environmental problems; they are the modern equivalents of salinised fields and denuded forests. The question is not whether humanity can innovate, but whether innovation will remain tied to ecological limits. Past societies remind us that intelligence without foresight is insufficient. *What undid them was not ignorance of nature but blindness to its thresholds.* The ruins of their fields and forests remind us that prosperity is never separate from ecology—it is ecology, temporarily arranged in our favour, and disastrously fragile when neglected.

War's Destructive Paradox: How Weaponry Undermined the States that Forged It

The history of warfare is often told as a story of progress: sharper blades, stronger armour, deadlier projectiles, each innovation expanding the reach of states and securing empires. Yet a subtler truth runs beneath the surface. The same ingenuity that delivered military advantage frequently destabilised the very societies that nurtured it. *Weaponry is not neutral; it reshapes power, institutions, and economies in ways that can corrode from within.* The paradox is that advances designed to protect states often accelerated their decline.

Take the chariot, one of antiquity's most celebrated military breakthroughs. Developed in the Near East around the second millennium BCE, it combined speed, shock, and prestige in a way that transformed warfare. Kings paraded them as symbols of divine sanction, and empires expanded under their power. Yet chariots were fragile machines, costly to maintain, and dependent on a specialised warrior elite. Their dominance entrenched hierarchical societies where power clustered in narrow classes of aristocratic fighters. As warfare spread and rivals adopted similar technology, the advantage diminished, leaving states with expensive militaries and rigid social structures vulnerable to upheaval. When infantry armed with cheaper, more versatile weapons eventually challenged the supremacy of the chariot, those states overly reliant on it found themselves unable to adapt. *The innovation that had once secured glory became a burden that sapped resilience.*

Gunpowder offers another illustration of how innovation destabilises its own progenitors. The Chinese pioneered its use not only for fireworks and ritual but also for war. By the 10th century, rudimentary gunpowder weapons were fielded, and by the 13th, the Mongols and others had spread the technology westward. Yet in China itself, the adoption of gunpowder coincided with turbulence. The centralised bureaucracy struggled to control the proliferation of firearms, and power increasingly shifted to local warlords and frontier groups who mastered the new weapons. Gunpowder promised protection but amplified fragmentation. A similar paradox unfolded in Europe centuries later. Cannon and firearms helped monarchs subdue rebellious nobles and build centralised states, yet they also necessitated vast expenditures. Maintaining gunpowder armies required taxation, debt, and bureaucracy on unprecedented scales. While some states managed the transition—France and England among them—others, like the smaller Italian city-states, were hollowed out by the financial burdens of perpetual militarisation. Weaponry granted advantage, but only at a cost many could not bear.

Consider the case of the Roman Empire. The legions were defined by disciplined infantry, but Rome was also a master at incorporating foreign weapons and tactics into its arsenal. This adaptability fuelled expansion, yet it also had corrosive effects. The empire's reliance on increasingly diverse mercenary forces, equipped with new styles of weaponry, eroded the cohesion that once defined the legions. Weapons themselves were not the cause of Rome's fall, but the social transformations they induced—empowering frontier recruits, diluting citizen soldiers, and fuelling reliance on external auxiliaries—contributed to the empire's eventual vulnerability. Here the paradox is clear: *a state that thrived on military innovation was destabilised by the very changes innovation produced within its armies.*

The destructive paradox intensifies in modern times. The First World War was a laboratory of weaponry: machine guns, poison gas, tanks, and heavy artillery. These inventions were heralded as solutions to stalemate, yet they deepened the carnage and prolonged the conflict. States emerged from the war not strengthened but exhausted, their societies fractured, economies destabilised, and political orders overturned. Russia's tsarist regime collapsed under the weight of mobilisation and losses, paving the way for revolution. Germany, whose military technology was among the most advanced, saw its monarchy fall and its economy crippled by both war and defeat. The war illustrated that *technological superiority in arms guarantees neither victory nor stability; it may instead magnify destruction beyond the state's capacity to withstand.*

Perhaps the most chilling expression of the paradox lies in nuclear weapons. Developed in the crucible of the Second World War, the atomic bomb gave the United States an unparalleled advantage in 1945. Yet within years, the Soviet Union had developed its own arsenal, inaugurating a nuclear arms race that destabilised the global order for decades. Far from securing safety, nuclear weapons institutionalised the condition of mutual vulnerability. The sheer destructiveness of the weapons rendered their use almost unthinkable, but their existence demanded costly military and political strategies to manage deterrence. States that pursued them often undermined their own stability: the Soviet Union, burdened by the economic and political weight of sustaining its military machine, collapsed in part under the pressures of the arms race. *The technology that promised ultimate security corroded the foundations of power itself.*

There is a philosophical dimension worth noting. Weaponry embodies both mastery and fragility. It grants control over enemies yet exposes states to new vulnerabilities: financial burdens, social strains, political dependencies. Each breakthrough disrupts balances—between rulers and ruled, centre and periphery, state and environment. The paradox is not accidental but structural: the logic of war rewards escalation, but escalation reshapes societies in destabilising ways. What begins as protection against external threats often becomes an internal crisis of governance, cohesion, or sustainability.

The paradox also invites reflection on the present. Cyberwarfare and autonomous weapons represent today's cutting edge. They promise precision, speed, and deterrence, yet they also destabilise in familiar patterns.

Cyberweapons proliferate rapidly, eroding the advantage of pioneers and creating vulnerabilities that can cripple the very infrastructures of the states that develop them. Autonomous weapons raise ethical and political questions that could fracture international norms and domestic trust. Just as the chariot entrenched fragile aristocracies and gunpowder bankrupted city-states, so too might digital or robotic warfare destabilise contemporary powers. The lesson is consistent: *innovation in arms is never confined to the battlefield; it reverberates through the fabric of society.*

What emerges, then, is a sober recognition. Societies have often imagined weaponry as tools they control, but history suggests weapons exert their own momentum, transforming the states that wield them. The paradox is not simply that weapons fail to secure lasting dominance, but that they alter power structures in ways that erode stability from within. States that embrace innovation without reckoning with its systemic consequences are caught in a cycle: they master weapons briefly, only to be mastered by the political, economic, and social upheavals those weapons unleash.

If there is a thread connecting the chariot, the cannon, the machine gun, and the nuclear bomb, it is this: *military brilliance contains the seeds of political fragility.* The very ingenuity that brings short-term dominance carries with it destabilising forces that no ruler, however powerful, has yet found a way to escape.

Forgetting as Collapse: How Knowledge Disappears When Systems Break

Civilisations rarely fall in a single dramatic moment. More often, they unravel quietly, not only through famine, invasion, or economic decline, but through something less visible and arguably more devastating: the erosion of memory. Knowledge, once lost, leaves a society not only weakened but disoriented, stripped of the intellectual inheritance that once sustained it. *The collapse of knowledge transmission systems has been as consequential as the collapse of armies or markets, for without memory, continuity falters and progress becomes fragile illusion.*

The ancient world provides stark reminders of how fragile the preservation of learning can be. The Mycenaean Greeks of the Late Bronze Age possessed a literate bureaucracy capable of maintaining palatial economies and extensive trade. Their administrative script, Linear B, allowed records of taxes, inventories, and transactions to be stored and consulted. Yet when the palaces fell in the twelfth century BCE, the scribal class collapsed with them. Writing disappeared entirely for centuries in what historians now call the Greek Dark Age. Oral tradition preserved fragments of myth and epic, but practical knowledge—the techniques of statecraft, large-scale construction, and international trade—vanished. Greece eventually reinvented writing through the borrowed Phoenician alphabet, but the gap severed institutional continuity. A civilisation that had known writing became, in effect, illiterate, forced to rebuild its intellectual infrastructure from scratch.

The case of the Maya is another illustration of how systems of knowledge can disintegrate when their carriers vanish. The Maya maintained a complex hieroglyphic script inscribed on monuments, codices, and ceramics. When the Spanish conquest began, thousands of Maya texts existed. Yet within a few generations, systematic burning by missionaries and the suppression of native scribal classes reduced this corpus to just four surviving codices. The result was not merely cultural destruction but the obliteration of astronomical, calendrical, and medical knowledge accumulated over centuries. Oral traditions preserved elements of Maya history, but the scientific sophistication encoded in the lost manuscripts was erased from practical use. It is telling that contemporary researchers often reconstruct Maya science not from living traditions but from painstaking archaeological and epigraphic work. What was once common knowledge to a literate elite is now the subject of speculative reconstruction.

Perhaps the most famous example of forgetting is tied to the destruction of libraries, above all the Library of Alexandria. While the popular imagination dramatises its burning as a single catastrophic event, historians suggest its decline was gradual, shaped by political instability, repeated conflict, and changing patronage. The larger point remains: when institutions responsible for storing and transmitting learning are weakened, the risk of irretrievable loss is high. Works of science, philosophy, and engineering known in antiquity are absent from the modern record not because they were never written but because the channels of preservation failed. The disappearance of technical treatises, such as those describing advanced Hellenistic automata or lost plays of classical theatre, reveals that *knowledge is not naturally cumulative; it must be actively sustained or it dissipates.*

Even societies that appeared secure in their intellectual traditions were vulnerable. The Roman Empire relied on an immense network of scribes, copyists, and libraries to maintain its literary and administrative records. Yet as the Western Empire fractured, literacy narrowed to a clerical elite within monasteries. Vast swathes of Roman law, engineering, and scientific theory vanished because no one copied them. The survival of classical texts today is less the result of an unbroken chain than of extraordinary contingency: the chance that a single manuscript was deemed worth preserving by a medieval monk, or that a copy escaped fire, damp, or neglect. The rediscovery of Aristotle's works in the twelfth century, preserved through Arabic and Byzantine transmission, transformed medieval European thought—yet it also underscored how close that knowledge had come to disappearing entirely.

Knowledge loss is not only a story of ancient civilisations. Early modern Europe itself provides an instructive case. During the English Civil War, many records of parish life, local law, and scientific experimentation were destroyed in the turmoil. Similarly, in China, the Qin dynasty's infamous "burning of books and burying of scholars" in the third century BCE is often cited as an act of deliberate epistemic violence: in seeking to consolidate authority, the state suppressed diverse traditions of thought. Though some texts survived, the erasure reshaped Chinese intellectual history by privileging certain schools over others. *Forgetting can be orchestrated, not only accidental; it may serve the interests of power but leave lasting damage to collective resilience.*

One must also consider how forgetting operates less dramatically, through neglect rather than fire. Agricultural techniques, for instance, often perish when they no longer align with prevailing economic or political systems. In the Andean highlands, pre-Columbian societies engineered sophisticated terracing and irrigation systems that sustained large populations in difficult climates. After the disruptions of conquest and colonial administration, much of this expertise disappeared. Only recently have archaeologists and agronomists begun to rediscover and reapply these techniques in response to climate stress. Here, forgetting was not an act of sudden destruction but of gradual disuse—a loss that impoverished communities facing environmental challenges that earlier generations had already solved.

The phenomenon reveals something profound about the nature of human knowledge. It is rarely the property of isolated individuals; it is embedded in systems—institutions, traditions, scripts, crafts, apprenticeships. When those systems are disrupted, knowledge cannot simply hover in the air awaiting rediscovery. *It decays like an untended field.* Oral transmission can preserve myth or ritual for astonishingly long periods, but technical knowledge often requires continual reinforcement and practice. To build a bridge, to cast a bronze, or to calculate an astronomical cycle is not merely to recall information but to enact skills that fade without repetition.

Modern society is not exempt from this risk. The digital age offers extraordinary tools for storage and retrieval, but it also introduces new vulnerabilities. Vast amounts of knowledge exist only in digital formats, dependent on fragile infrastructures. A failure of archiving, a collapse of platforms, or the obsolescence of file formats could render entire domains of information inaccessible. Already, historians of the late twentieth century note that digital records decay faster than paper archives; magnetic tapes demagnetise, CDs degrade, servers shut down. Future historians may know less about the early internet than we know about medieval Europe, not for lack of data but because the transmission systems are unstable. *We imagine that abundance secures permanence, but without continuity of stewardship, abundance accelerates forgetting.*

There is also the question of social will. Knowledge survives not merely through technical means but through cultural commitment to preservation. When societies devalue certain forms of expertise—whether traditional ecological knowledge, minority languages, or manual crafts—the infrastructure of transmission collapses. Consider the erosion of indigenous knowledge systems under colonial pressures. Languages vanish, and with them disappear taxonomies of plants, animals, and medicines refined over centuries. The cost is not only cultural but practical: lost insights into biodiversity and sustainability at a moment when they might be crucial. Forgetting, in such cases, is not a relic of ancient collapse but an active process of marginalisation still unfolding.

The larger lesson is stark. Civilisations that imagine knowledge to be cumulative and secure misunderstand its fragility. Libraries burn, scribes perish, digital archives fail, traditions erode under conquest or neglect. The collapse of memory is often less visible than the collapse of walls, but its consequences are deeper, for it leaves societies not only weakened but unable to learn from their past. The Roman aqueduct crumbles, but the greater tragedy is when the method of its construction is no longer known. A field lies barren, but more devastating is when the knowledge of how to restore its fertility has been lost.

What history shows is that the continuity of knowledge requires redundancy, diversity, and care. Redundancy ensures that learning is not entrusted to a single medium or institution. Diversity guarantees that no monopoly can erase or distort collective memory entirely. Care means recognising that the work of preservation is as vital as the work of innovation. Societies that fail to invest in these principles are not merely vulnerable to collapse; they are vulnerable to amnesia, a form of collapse that leaves them unable to even name what has been lost. *Forgetting is not only a symptom of decline; it is a cause, one of the most insidious forces that turns flourishing into fragility.*

Chapter 12
The Forgotten Mental Framework

Human progress seldom unfolds as a straightforward narrative of tools shaping thought. More often, it is thought shaping tools. The decisive leaps in civilisation did not spring from the first smelting of iron, the initial harnessing of steam, or the early flicker of electricity. They emerged earlier, almost invisibly, as conceptual breakthroughs: the capacity to discern unseen connections, to model possibilities, to reconceive limitations as openings. Humanity's richest resource has rarely been material—it has always been mental. It is this invisible scaffolding of intellect, the architecture of thought, that forms the spine of this chapter.

Take the bow and arrow. Its significance lies less in wood, sinew, or stone than in the conceptual leap: the idea of storing energy and releasing it at will. Similarly, the wheel was not merely a round stone repurposed; it was the recognition that rolling reduces friction and permits the transport of heavier loads with less effort. Both required physical construction, yes—but the greater advance was cognitive. One must conceive a principle before embodying it in matter. Raw material alone is inert; imagination animates it. Too often, histories of progress privilege artefacts, obscuring the crucial precondition: thought made tangible.

Imagination itself functions as a technology. It is a tool that shapes the horizon of possibility, guiding the allocation of effort and resources. Consider double-entry bookkeeping in Renaissance Italy. No new machine, no novel material, yet merchants gained a cognitive instrument for tracking debts, credits, and investments across continents. It proved as transformative as compass or cannon in shaping commerce and exploration. Likewise, the concept of zero in Indian mathematics, transmitted via the Islamic world to Europe, was not a tangible device but a framework that revolutionised calculation, algebra, and ultimately computation. To view history solely through the lens of artefacts is to miss that invisible engines often drive visible change.

Constraint, far from stifling ingenuity, can sharpen it. When resources abound, the mind often drifts; when scarce, it leaps. Polynesian navigators exemplify this: crossing vast oceans without compasses or maps, relying on stars, wave patterns, bird flight, and cloud formations. Their achievement arose not despite scarcity but because of it. Similar patterns recur through history. During the Second World War, shortages compelled British and German engineers to improvise with materials, producing synthetic rubber, plastics, and fuels that reshaped postwar industry. Tokugawa-era Japan, starved of raw metals, advanced metallurgy and recycling techniques. In architecture, brilliance often sprang from limitation: mud-brick Mesopotamian structures, Gothic stone vaults, compressed-earth contemporary designs. In each case, scarcity demanded refinement of vision.

Innovation need not wait for material invention. Intellectual shifts can precede and outweigh physical discovery. Darwin's theory of natural selection required no laboratory apparatus, yet it transformed biology and philosophy. Adam Smith's "invisible hand" was a conceptual device reshaping economic life. Einstein's relativity began as thought experiment, not mechanical contraption. In each instance, the primary technology was mental; instruments and applications followed, often decades later.

This emphasis has contemporary relevance. Today, technological capacity—quantum computing, genetic engineering, artificial intelligence—surges, yet our challenges are more often conceptual than technical. Climate change is less a problem of tools than of reframing growth, consumption, and sustainability. Technologies exist; the frameworks for integrating them socially and politically lag behind. Likewise, AI debates reveal that our true task is not building machines, but imagining ethical, economic, and cultural models capable of governing their deployment. Our scarcity is not of artefacts, but of imagination commensurate with complexity.

The mental framework this chapter seeks to recover is simple yet profound: thought itself is a technology. It is a tool as real and consequential as hammer, plough, or algorithm. The capacity to imagine alternatives, model constraints, and synthesise across domains underpins three key principles:

- Conceptual leaps precede and drive material progress.
- Scarcity sharpens rather than stifles ingenuity.
- Intellectual tools—mental models, systems of thought, and frames of reference—are as decisive as physical invention.

These principles are more than historical insight; they are practical guides. In an age of complexity, with limited resources and uncertain horizons, leverage lies less in accumulating tools than in refining the frameworks that direct them. A constrained business may spark strategic creativity; a research team with limited data may produce models more elegant than one awash in information; a policymaker with scarce political capital may achieve impact through reframing rather than brute force.

The forgotten mental framework is not relic but imperative. Material technologies rise and fall, often within decades. Thought, however, multiplies as it is used. By studying how past civilisations harnessed it, and how they faltered when they neglected it, we recover a tool for our present: the ability to reframe, model, and imagine. Progress has never been solely material; it rests upon the intellectual scaffolding that shapes material into meaning. To reclaim this framework is to wield the most enduring tool humanity possesses.

Imagination as Technology

The decisive breakthroughs in human history were not born from the discovery of materials but from the ability to reimagine what those materials could do. Bronze existed long before the Bronze Age, just as iron lay dormant in the earth for millennia before it reshaped empires. What transformed them from passive elements into instruments of civilisation was not their presence but the mental leap that saw possibility where others saw only stone and soil. *The true technology of humanity has always been imagination—the capacity to model, to anticipate, to bridge the gap between the visible and the conceivable.*

Take the example of fire. Early hominins encountered fire for hundreds of thousands of years as a destructive force. What made it a tool was the conceptual leap of seeing that it could be harnessed, contained, and applied for cooking, warmth, and protection. No new material appeared on the earth that had not been there before; what changed was the framework of thought. Fire became the earliest transformative technology precisely because it was reinterpreted by imagination. The same pattern repeats with other breakthroughs: the wheel required the insight that rolling reduces friction, the lever depended on grasping the principle of mechanical advantage, and agriculture emerged from the conceptualisation of time, seasonality, and growth cycles rather than from the simple act of planting seeds. Each case demonstrates that the raw materials of progress are cognitive before they are physical.

This recognition reframes the way we view history. We are often tempted to narrate the past as a succession of material discoveries: the Stone Age, the Bronze Age, the Iron Age, the Industrial Age, the Digital Age. Yet the labels are misleading, for they suggest that material substance alone defined progress. In truth, stone was used differently across millennia, bronze enabled different cultural expressions depending on social organisation, and iron could be a tool of agriculture or of conquest depending on the imagination that directed it. What distinguishes epochs is not merely what humans had in their hands but what they carried in their minds.

One of the clearest illustrations of this is the invention of writing. Clay tablets in Mesopotamia or papyrus in Egypt did not in themselves give rise to written culture. The breakthrough was the recognition that language, transient and spoken, could be represented symbolically and fixed into external memory. This leap of abstraction transformed administration, law, and culture, permitting societies to scale far beyond the limits of oral tradition. Writing was less an artefact than a system of thought externalised, and its consequences echo to this day in how we conceptualise knowledge itself. *It was not the stylus or the tablet that mattered, but the imaginative step that turned marks into meaning.*

Imagination as technology also explains why societies with similar resources developed along such different trajectories. The peoples of Mesoamerica built cities, calendars, and astronomical systems with stone tools alone, while their counterparts in the Near East harnessed bronze and iron. Both achieved complexity, not because of the superiority of one material over another, but because of the conceptual frameworks they employed. Similarly, the Inca engineered a vast empire across the Andes without wheels, iron, or written script, yet through imaginative use of quipus (knotted cords for record-keeping) and terraced farming, they sustained millions of people in difficult terrain. Their ingenuity lay not in lacking materials but in reconfiguring the possibilities of what they had.

What makes imagination particularly powerful as a form of technology is its recursive quality: it does not diminish with use, but expands. The capacity to form conceptual models generates new questions, which in turn inspire further models. Mathematics illustrates this dynamic vividly. The introduction of zero and the decimal place-value system in India was not a material invention but a cognitive leap. It allowed for efficient calculation, advanced algebra, and eventually the foundations of modern computation. The material tools of calculation—abaci, counting rods, later slide rules and computers—would come and go, but the conceptual framework outlasted them all. Indeed, the computers of the twenty-first century still rely on the binary system, itself another act of conceptual simplification: reducing the infinite complexity of information to two states, one and zero.

This perspective also sheds light on the fragility of progress. When imagination is constrained, suppressed, or forgotten, societies can stagnate even amidst abundance. The collapse of complex polities often coincides not only with resource depletion or external invasion but with a failure of conceptual adaptability. Ancient Egypt, for instance, displayed remarkable stability but at times became rigid in its worldview, resisting innovation in military or administrative models even when confronted with more adaptive neighbours. By contrast, smaller or materially poorer societies that fostered imaginative thinking often punched above their weight. Athens, with few natural resources, gave the world enduring models of philosophy, science, and democratic governance—not because it had more, but because it thought differently about what it had.

It is tempting to imagine that the modern world, saturated with advanced machinery, has outgrown this dependence on imagination. Yet contemporary challenges reveal the opposite. Climate change, for example, is less a failure of technical tools than of conceptual models. Renewable energy technologies exist, but their integration depends on reimagining economics, consumption, and long-term responsibility. Artificial intelligence advances rapidly in hardware and algorithms, yet the decisive questions are imaginative: how should it be governed, what role should it play in education, healthcare, or justice, and what ethical frameworks are adequate to the pace of change? The technology that matters most here is not the chip or the server but the imagination to see beyond immediate gain and into systemic consequence. *The primary bottleneck is not invention but vision.*

Philosophers have long recognised the technological character of imagination. Plato's allegory of the cave is itself a model of cognitive technology: a thought experiment to reframe perception and reality. Kant described imagination as the faculty that synthesises sense and concept, making knowledge possible. More recently, Hannah Arendt emphasised imagination as central to political life—the ability to "think from the standpoint of others." Across these traditions, imagination is not merely fanciful or decorative but structural: it makes possible new configurations of reality that can then be acted upon.

This reframing also forces us to reconsider the meaning of innovation in contemporary culture. Too often, innovation is equated with product development, gadgets, or material novelty. But the deepest innovations arise from conceptual breakthroughs. Double-entry bookkeeping, once dismissed as clerical technique, underpinned the rise of capitalism and global trade. The germ theory of disease, a conceptual reordering of how illness was understood, saved more lives than any single medical device. These were not "inventions" in the narrow sense but imaginative technologies—mental architectures that reorganised reality.

To cultivate imagination as technology in the present requires resisting the seduction of material fetishism. It is easier to measure progress in terms of devices manufactured, patents filed, or infrastructures built. But the more important metric is the fertility of conceptual frameworks. Educational systems that prize memorisation over creative reasoning, or economies that reward short-term exploitation over imaginative resilience, risk impoverishing the very resource that drives sustainable progress. In contrast, societies and organisations that protect spaces for speculative thinking, interdisciplinary exchange, and even play are investing in the most renewable technology of all.

The paradox of imagination is that it is both universal and unevenly harnessed. Every human possesses it, yet not every culture or institution fosters it equally. This raises a practical question: how can imagination be deliberately cultivated as a tool? The historical record suggests that conditions of dialogue, cross-cultural contact, and constraint tend to stimulate it. The Greek city-states, the Abbasid translation movement in Baghdad, the European Renaissance—all were crucibles where diverse perspectives interacted, sparking conceptual leaps. Today, globalisation provides a similar opportunity, though it also risks homogenisation if not paired with a respect for plural imagination. *To treat imagination as technology is to see it as something that can be designed for, nurtured, and refined, not merely left to chance.*

Ultimately, the lesson of history is clear: materials matter, but they are inert without the frameworks that make them meaningful. The imagination transforms clay into law, marks into memory, numbers into algorithms, scarcity into innovation. Progress is not dictated by what lies beneath the earth but by what emerges in the mind. If we forget this, we risk mistaking artefacts for causes and abundance for advancement. If we remember it, we can approach the challenges of our own time with the recognition that the most powerful technology we possess is not external but internal—the capacity to imagine differently, and thereby to make the world anew.

Innovation Without Invention

The most transformative advances in human history often involved no new object at all. They arose from ways of seeing, thinking, and organising experience that redefined what was possible. The double-entry bookkeeping system of the late medieval period was nothing more than a method of writing numbers on a page, yet it reorganised commerce, finance, and ultimately capitalism. The discovery of perspective in Renaissance art did not produce a new material but a new way of imagining space, which in turn influenced architecture, cartography, and even scientific observation. *Innovation without invention reminds us that thought, not artefact, is the deepest driver of progress.*

To understand this distinction, it is necessary to separate invention—the creation of a tangible artefact or device—from innovation, which involves the application of a new framework or model that alters how people think and act. Invention tends to be visible and celebrated: the printing press, the steam engine, the aeroplane. Innovation in the deeper sense is subtler. It may manifest as a shift in how a society understands value, time, or organisation. These shifts are not always accompanied by a new machine, but they often prepare the ground for entire waves of invention. *Mental models are not accessories to progress; they are its preconditions.*

Consider the concept of zero. Introduced and elaborated in India and transmitted through the Islamic world to Europe, zero was not an object but an idea, a conceptual placeholder that enabled new forms of calculation. Without it, modern mathematics, physics, and computer science would be inconceivable. The invention of physical calculating devices or electronic computers centuries later was derivative of this earlier conceptual innovation. The artefacts followed the model, not the other way round.

The same can be said of timekeeping. The invention of the mechanical clock in the fourteenth century was important, but the greater innovation was the reconceptualisation of time itself. For most of human history, time was seasonal, fluid, and tied to natural cycles. The spread of clocks imposed a new framework: time as segmented, measurable, and exchangeable. This shift reshaped labour, productivity, and social organisation. By the eighteenth century, it underpinned the factory system and the very concept of efficiency. Here again, the tool mattered less than the mental model it embodied.

When innovation is understood as a shift in thought, its significance extends far beyond technology. The Protestant Reformation, for example, was not born of a new invention but of a reorientation of how authority and interpretation were understood. The idea that individuals could read scripture directly destabilised centuries of ecclesiastical control. Printing amplified this transformation, but the decisive innovation was intellectual: the decentralisation of religious authority. Similarly, the scientific revolution was less about telescopes and microscopes than about adopting a new model of inquiry—systematic observation, hypothesis, and verification—that allowed those instruments to be used in radically different ways.

What makes mental models so powerful is their ability to reconfigure entire systems of practice. Double-entry bookkeeping did not simply help merchants keep cleaner accounts. It allowed them to visualise profit and loss, to track capital across time, to conceptualise the future as a space of calculable risk. That shift in thought changed how businesses, and eventually whole economies, functioned. Today, financial instruments such as insurance, credit, and derivatives trace their lineage to this conceptual innovation. The physical tools of record—ledgers, pens, later computers—are secondary to the mental model that gives them coherence.

This primacy of thought challenges the way progress is commonly narrated. Societies are often judged by their artefacts: pyramids, aqueducts, railways, or skyscrapers. Yet such monuments can obscure the subtler but more enduring shifts in understanding. Rome's roads were remarkable, but its legal system and ideas of citizenship left deeper legacies. The Enlightenment is remembered for scientific discovery, but its more profound contribution was the model of universal rights and reasoned debate. These were not inventions in the material sense, yet they altered the trajectory of civilisation more than any single device could.

It is also important to recognise that physical invention without supporting mental models rarely scales effectively. Gunpowder, invented in China, initially served ritual and decorative purposes. It was only when Europeans developed new models of warfare and state organisation that it became the foundation of modern military power. Similarly, the wheel existed for centuries before it was used for transport, because the conceptual leap required to apply it to vehicles came slowly. Artefacts without models can remain curiosities; models without artefacts can reshape entire societies.

The digital age provides abundant contemporary examples of this distinction. The core innovation of the internet is not the fibre optic cable or the server, but the idea of a network of information freely accessible across distance. Email was not a new artefact—it was a reconfiguration of how communication could be imagined: instant, asynchronous, and decoupled from physical presence. Likewise, blockchain is not primarily about cryptographic code but about a conceptual shift in how trust and verification can be organised without centralised authority. These examples demonstrate that what matters most is the model; the artefacts are expressions of that model.

Philosophically, this underscores a deeper truth about human cognition. Artefacts are finite, bounded by their material conditions. Mental models are infinite in their generativity. They can be recombined, abstracted, and reapplied across contexts. The idea of evolution by natural selection, for example, has influenced not only biology but also economics, sociology, and computer science. The model becomes a tool that travels, shaping diverse domains without requiring new physical inventions. *The portability of thought is the true measure of its technological power.*

At the same time, innovation without invention highlights a vulnerability. Because mental models are intangible, they are also fragile. They can be lost, suppressed, or forgotten more easily than artefacts. The decline of classical knowledge in post-Roman Europe illustrates this. The texts of Greek philosophy and science were preserved in libraries but inaccessible to much of the population. The mental frameworks they contained—rational inquiry, logical argument, mathematical reasoning—waned, leading to centuries where invention continued but innovation faltered. Artefacts endured, but without their underlying models, their full potential could not be realised.

This fragility is echoed in the modern world. Organisations often obsess over patents, prototypes, and products while neglecting the cultivation of conceptual innovation. A company may invest millions in research and development but fail to ask whether its mental models of markets, users, or sustainability remain valid. The recent crises in global finance, for instance, were not due to a lack of technical sophistication but to flawed conceptual models of risk, value, and growth. When models fail, no amount of artefactual invention can compensate.

Recognising the primacy of thought over artefact offers practical lessons. It suggests that education should prioritise not only technical skill but also the cultivation of flexible conceptual frameworks. It urges policymakers to focus less on physical infrastructure alone and more on the ideas that make such infrastructure meaningful. It challenges innovators to ask not just what they can build, but what ways of thinking they are enabling or obstructing. Above all, it reminds us that progress is measured not by the proliferation of devices but by the expansion of human understanding.

The distinction between innovation and invention also helps explain why scarcity often catalyses ingenuity. Material abundance can encourage complacency, but constraint forces the imagination to reconfigure possibilities. The Polynesians who navigated the Pacific without compasses or sextants did so through conceptual models of stars, currents, and bird flight. Their innovation lay in thought, not in artefact. Today, the most pressing challenges—climate change, inequality, global health—demand precisely this kind of mental reconfiguration. The artefacts exist or can be developed; the barrier is the model through which they are understood and deployed.

When we speak of human progress, then, it is less accurate to say that it is driven by invention than by the interplay of innovation and thought. Artefacts are the crystallisation of ideas, but it is the idea that precedes and outlasts the material form. A steam engine rusts, a factory crumbles, but the concept of mechanical power endures and multiplies across new contexts. To grasp this is to see history not as a museum of objects but as a living archive of imagination. It is to understand that what most distinguishes humanity is not what we build but how we think about what can be built. And it is to recognise that the greatest frontiers ahead of us will not be opened by machines alone, but by the conceptual leaps that show us how to use them.

Constraint as Catalyst

Scarcity is often imagined as an obstacle to human progress, a brake on ambition and achievement. Yet history consistently reveals the opposite: limitation, whether of resources, environment, or opportunity, has served as a prod to ingenuity. When abundance cushions societies, they tend to stagnate, repeating inherited patterns. It is in the tightening of circumstances that imagination sharpens, disciplines strengthen, and unconventional solutions emerge. *Constraint is not the enemy of human creativity—it is the furnace in which it is forged.*

One of the clearest demonstrations of this principle comes from ancient civilisations in arid landscapes. Mesopotamia, wedged between the unpredictable floods of the Tigris and Euphrates, had no fertile plains in the sense enjoyed by the Nile valley. Instead, it faced salinisation, erratic harvests, and a precarious dependence on irrigation. These difficulties did not diminish its development; they demanded systemic innovation. The first codified legal systems, methods of urban administration, and organised agriculture all took shape in response to these challenges. Here, scarcity of predictability bred the intellectual tools of governance. Egypt, by contrast, with its relatively dependable Nile, produced wonders of art and architecture but displayed slower shifts in political and economic organisation. Predictability conferred stability, but adversity forced invention.

The history of seafaring demonstrates a similar logic. The Polynesians, inhabiting small islands scattered across the Pacific, faced extreme scarcity of land and resources. Their solution was not resignation but an extraordinary mastery of navigation. Without compasses or charts, they devised mental models of ocean swells, bird flight patterns, and star movements to traverse thousands of miles of open sea. European sailors, centuries later, marvelled at this achievement. The Polynesians' resource constraints pushed them to cultivate the mind as the ultimate instrument. Scarcity of material tools spurred abundance of cognitive technique.

Modern engineering frequently confirms that constraint accelerates, rather than impedes, creativity. The Manhattan Project is often remembered for its vast resources, but in practice, its breakthroughs depended on severe constraints: lack of time, uncertainty of outcome, and existential stakes. These forced interdisciplinary collaboration and accelerated experimentation, setting a precedent for "big science" that persists today. Similarly, the space race was defined by limitations—tiny payloads, limited fuel, fragile life-support systems. These boundaries pushed scientists to miniaturise electronics, refine materials, and create computational systems far more advanced than any peacetime context would have required. Each obstacle became a generative force.

Constraint also reveals its catalytic role in the arts. The sonnet, with its rigid 14-line structure, did not suffocate poets but gave rise to some of the most enduring works of literature. Japanese haiku, confined to a 17-syllable pattern, distilled perception to its essence. In music, the limited harmonic rules of the Baroque period did not produce monotony but a wealth of variation. *Form, when bounded, often becomes richer, because constraint forces attention to the quality of each choice.* Unlimited freedom risks dilution; boundedness sharpens focus.

This interplay between scarcity and ingenuity can be seen at the societal level in times of crisis. Britain during the Second World War endured rationing and resource shortages that forced innovations in nutrition, logistics, and urban planning. From powdered eggs to radar, from urban allotments to advances in information management, constraint fostered resilience and new systems of organisation. Contrast this with periods of resource glut, such as the oil booms of the 1970s in certain states, which produced wealth but also waste, inefficiency, and dependence. Abundance often seduces societies into complacency; constraint obliges foresight.

Yet it would be simplistic to suggest that constraint is universally positive. Limitations can, in some cases, extinguish possibility. The collapse of Easter Island's ecosystem is a case where scarcity, exacerbated by overuse, led not to ingenuity but to social breakdown. Famines have often devastated rather than invigorated communities. The critical distinction lies in whether constraint is perceived and addressed through adaptive frameworks. When societies cultivate mental flexibility, scarcity pushes them forward; when they cling to rigid models, scarcity overwhelms. *Constraint is not inherently a teacher—it becomes one only when the mind is prepared to learn.*

The digital revolution offers a compelling modern case of constraint-driven creativity. Early computers were vast machines with minuscule memory. These limitations required programmers to develop languages and algorithms that were extraordinarily efficient, elegant in their economy of command. Today's abundance of storage and processing power has, paradoxically, encouraged sloppy code and bloated systems. Indeed, some of the most effective software still comes from contexts where constraints—battery life, bandwidth, or security—force developers to design with care. The smartphone, the most transformative device of the 21st century, was born out of the demand to concentrate multiple functions into a small, resource-limited device. Constraint guided the design, and its ubiquity is testament to the power of those pressures.

There is also a deeper philosophical resonance in the relationship between constraint and creativity. Human thought itself operates under limitation—of memory, of attention, of lifespan. Were our minds infinite, they might never need to prioritise, simplify, or abstract. It is precisely because we cannot hold everything that we learn to invent categories, metaphors, and systems. Cognition is the art of working within limits. In this sense, every act of human understanding mirrors the principle that scarcity is generative. The great ideas of history—democracy, algebra, relativity—arose because individuals had to wrestle with constraints, whether political, linguistic, or perceptual.

The lessons of constraint are particularly urgent today. In a world facing ecological and economic limits, the temptation is to treat scarcity as catastrophe. Yet history suggests it may be the necessary spark for a new phase of human creativity. Renewable energy, circular economies, and sustainable agriculture are all born of recognition that infinite growth within finite systems is impossible. Rather than treating these limits as prohibitions, societies that embrace them as design parameters may discover innovations more profound than those produced in times of plenty. Constraint forces the question: how much can be done with less? The answers are often surprising, sometimes revolutionary.

Understanding constraint as catalyst also challenges the modern fetish for abundance in personal and organisational life. Companies with limitless budgets often generate mediocre products because they lack pressure to refine. In contrast, start-ups operating under tight constraints are forced to innovate nimbly, cut through excess, and identify what truly matters. In education, too, resource-limited schools often develop creative pedagogies, relying on community and improvisation, while well-funded institutions risk bureaucratic inertia. *Necessity, when engaged with creatively, breeds depth and focus in a way abundance rarely does.*

The principle extends to individuals. Writers often find inspiration by setting arbitrary restrictions: composing within a strict word count, choosing a single theme, or limiting vocabulary. Athletes train under simulated adversity—altitude chambers, weighted vests—to sharpen their capabilities when conditions are easier. Entrepreneurs adopt lean methods not because abundance is unavailable, but because constraint clarifies priorities. In each case, limitation becomes a deliberate tool to ignite ingenuity.

Seen through this lens, scarcity is less a deficit than a field of possibility. It narrows options but intensifies imagination. It compels discipline, resilience, and resourcefulness. The critical insight is that *human ingenuity is not the overcoming of constraint, but its transformation into opportunity*. The walls do not shrink the space of creativity; they give it definition. History's great leaps—from Mesopotamian irrigation to Polynesian navigation, from wartime innovation to digital efficiency—remind us that progress has rarely been born in the comfort of plenty. It has emerged, again and again, in the crucible of constraint.

A Mindset for Today

The paradox of modernity is that, despite the immense abundance of information, energy, and technology at our disposal, individuals and institutions alike often feel paralysed in the face of complexity. Decision-making is swamped by data, organisations stall under bureaucratic weight, and individuals are left grasping for clarity in lives saturated with choice. Yet when we examine the arc of human progress, we find that the most profound breakthroughs were rarely the result of endless resources or infinite options. They were products of constraint, of disciplined imagination, and of the deliberate cultivation of mental frameworks that could distil chaos into coherence. The mindset that enabled ancient navigators to cross vast oceans without instruments, or early mathematicians to conceive of zero before it had any physical counterpart, is precisely the mindset most needed today. *Modern abundance has not eliminated the relevance of ancient mental tools; it has amplified their necessity.*

At the heart of this mindset is the recognition that thought precedes tool. The modern world is enamoured with devices, platforms, and technologies, but their true power lies not in their material construction but in the mental models they embody. The abacus did not make calculation possible; the conceptual leap of representing numbers symbolically did. The printing press was not born merely of mechanical ingenuity but of the conviction that ideas could—and should—circulate beyond the bounds of oral tradition and manuscript culture. When we confuse the artefact for the advance, we risk mistaking accumulation for innovation. A mindset for today requires us to invert this view: to treat technologies as temporary vessels for enduring intellectual strategies, and to focus our efforts on sharpening the frameworks that generate and guide them.

Consider the challenge of climate change. Much of the discourse centres on technological solutions: renewable energy, carbon capture, geoengineering. These are vital, but they are expressions of a deeper requirement: a shift in mental models from linear growth to systems thinking, from extraction to regeneration. Civilisations that survived environmental pressures in the past did so not by outbuilding their peers but by rethinking their relationship to scarcity. The lesson is clear—before we can innovate responsibly, we must cultivate the capacity to reconceptualise problems at the level of thought. This is not a matter of having more tools but of training the imagination to treat limits as design constraints rather than dead ends. *Frameworks, not factories, determine whether societies collapse or endure.*

A mindset fit for contemporary complexity also requires comfort with paradox. Ancient thinkers often held contradictory models simultaneously, using each as a lens rather than a claim of absolute truth. Greek philosophers debated whether reality was continuous or discrete; Indian logicians explored the coexistence of multiple, seemingly opposing truths. These frameworks did not resolve paradoxes by erasing them but used them as intellectual scaffolding for deeper inquiry. Today's challenges—balancing economic growth with ecological stability, or privacy with connectivity—demand precisely this capacity. Linear, binary models falter under the weight of interdependence. The mindset we inherit from ancient traditions teaches us that clarity often comes not from choosing one side of a paradox but from inhabiting the tension productively.

Scarcity, too, must be reinterpreted. For much of modern industrial life, scarcity was treated as the ultimate threat, something to be eliminated by expansion, efficiency, and conquest. Yet history shows that scarcity has been one of humanity's most reliable teachers. It concentrated Polynesian navigators on reading subtle environmental signals; it compelled wartime Britain to invent new logistical systems; it pushed early computer scientists to write code of elegant efficiency. In personal and organisational life today, scarcity often takes the form of time, attention, or energy rather than material resources. The response is the same: constraints can sharpen purpose. By accepting rather than fleeing from limits, we learn to identify priorities, strip away excess, and channel creativity into what matters. *Scarcity is not merely an economic condition but a psychological catalyst, and its intelligent embrace distinguishes resilience from collapse.*

This perspective has profound practical applications. For professionals, the principle of mental primacy over material abundance suggests that true competitive advantage lies not in accumulating more tools or data but in refining the frameworks through which they are interpreted. The financial analyst who understands systemic risk, rather than simply amassing datasets, is better equipped for volatility. The medical researcher who rethinks disease models, rather than chasing incremental pharmaceutical advances, may open entirely new avenues of treatment. For organisations, the lesson is that cultivating a culture of adaptive thought is more valuable than investing endlessly in infrastructure. *The best preparation for an uncertain future is not stockpiling but sense-making.*

Individuals, too, benefit from adopting this mindset in everyday life. A professional overwhelmed by competing demands can treat time constraints not as a barrier but as an ally in clarity: the limitation compels the ruthless identification of what truly matters. A student drowning in information can learn from ancient rhetorical training, which prized the art of memory, synthesis, and distillation over mere recall. Even in personal decision-making, the principle of inhabiting paradox is vital. Choosing a career path, for instance, is often framed as a binary between passion and pragmatism. A mindset shaped by ancient lessons resists this false dichotomy and instead explores how apparently competing values can be integrated, reframed, or sequenced over time.

There is also an ethical dimension. Ancient thought consistently placed imagination in service of community rather than individual advancement. The great legal codes, philosophical systems, and navigation techniques were collective tools, designed to preserve and transmit knowledge for the survival of the group. In a world increasingly driven by individual consumption and personal optimisation, reclaiming this orientation has never been more urgent. A mindset for today must balance creativity with responsibility, recognising that frameworks of thought shape not only personal outcomes but collective futures. To innovate without regard for the social and ecological systems that sustain us is to misuse the very mental powers that once ensured our survival.

Some may argue that today's abundance and complexity render ancient strategies obsolete, that the velocity of technological change demands novel cognitive tools. Yet when examined closely, the inverse appears truer. The more intricate the system, the more valuable the ability to simplify without distortion. The faster the pace of change, the more important the capacity to pause, question, and reconceptualise. The more abundant the options, the more necessary the discipline of constraint. The mental frameworks that shaped humanity before machines are not artefacts of a slower age; they are the enduring architecture of adaptability itself.

To live and think with this mindset is to cultivate a few enduring practices: prioritising conceptual clarity over accumulation, treating scarcity as a guide rather than an adversary, embracing paradox as a source of strength, and orienting imagination towards shared futures. These are not antiquarian curiosities but vital skills for navigating modern life. The executive grappling with organisational complexity, the policymaker confronting ecological limits, the individual striving for coherence in an overloaded world—all stand to benefit from these lessons. What unites them is the recognition that the most powerful technology has always been the human mind, and that its refinement depends less on abundance than on discipline.

The forgotten frameworks of ancient thought remind us that progress is not the product of tools alone. It is the cultivation of ways of seeing, thinking, and reimagining that persist across time. By re-engaging these mindsets with the full force of modern knowledge, we stand not merely to manage complexity but to turn it into a stage for new kinds of clarity and invention. The machinery around us will continue to evolve, but its true potential will always depend on the frameworks we bring to it. The task, then, is neither nostalgic nor utopian. It is to remember what scarcity, imagination, and paradox have always taught: that the mind, disciplined by limits and expanded by vision, remains humanity's most enduring instrument.

About the Author

Rowan Mercer is a writer and researcher of the history of ideas and material culture, focused on how people solve hard problems with limited tools. Mercer brings together archaeology, philosophy of science, and cognitive psychology to reconstruct the reasoning inside ancient navigation, architecture, agriculture, and warfare. Drawing on close readings of texts and hands-on study of sites and artifacts, Mercer writes in clear, precise prose that makes complex thinking legible and useful. The guiding conviction is simple: innovation begins in the mind, long before it becomes metal or stone. Mercer's work invites readers to look past modern gadgetry and recover practical mental models—observation, disciplined inference, and design under constraint—that sharpen decision-making in today's resource-strained world.

www.ingramcontent.com/pod-product-compliance
Lightning Source LLC
Chambersburg PA
CBHW031337230926
51848CB00002B/279
9789371774093